MARINE BIOLOGY

FISH ECOLOGY

MARINE BIOLOGY

Additional books in this series can be found on Nova's website under the Series tab.

Additional E-books in this series can be found on Nova's website under the E-books tab.

MARINE BIOLOGY

FISH ECOLOGY

SEAN P. DEMPSEY
EDITOR

Nova Science Publishers, Inc.
New York

For permission to use material from this book please contact us:
Telephone 631-231-7269; Fax 631-231-8175
Web Site: http://www.novapublishers.com

Additional color graphics may be available in the e-book version of this book.

Library of Congress Cataloging-in-Publication Data

Fish ecology / editors, Sean P. Dempsey.
p. cm.
Includes index.
ISBN 978-1-61324-282-7 (hardcover)
1. Fishes--Ecology. I. Dempsey, Sean P.
QL639.8.F55 2011
597--dc22
2011010129

Published by Nova Science Publishers, Inc. † New York

CONTENTS

Preface		**vii**
Chapter 1	Metabolic Rate of Marine Fish in Early Life and its Relationship to their Ecological Status *Irina Rudneva, Valentin Shaida*	**1**
Chapter 2	Antioxidant Defense in Marine Fish and its Relationship to their Ecological Status *Irina Rudneva*	**31**
Chapter 3	Does Habitat Affect the Genomic GC Content? A Lesson from Teleostean Fish: A Mini Review *Ankita Chaurasia, Erminia Uliano, Luisa Bernà, Claudio Agnisola and Giuseppe D'Onofrio*	**61**
Chapter 4	Genetic Structure of Masu Salmon, Oncorhynchus Masou *Shigeru Kitanishi and Toshiaki Yamamoto*	**81**
Chapter 5	Facultative Catadromy in the Freshwater Eel Genus Anguilla between Fresh Water and Sea Water Habitats *Naoko Chino and Takaomi Arai*	**99**
Chapter 6	Impacts of Anthropic Factors on Native Freshwater Fish in Brazilian Semiarid Region *Sathyabama Chellappa, Wallace Silva do Nascimento, Thiago Chellappa and Naithirithi T. Chellappa*	**115**
Chapter 7	Fish Ecology, Conservation Biology and New Insights from the Archaeological Evidence in the Beagle Channel (Tierra del Fuego, Argentina) *Atilio Francisco Zangrando and María Paz Martinoli*	**131**
Chapter 8	Ecology of Early Life-History Stages of Anadromous Shads *Eduardo Esteves*	**151**

Chapter 9 Heat Shock Proteins Modulate Signaling Pathways in Survival of Stressed Fish to Polluted Environments **173**
Ekambaram Padmini

Index **193**

PREFACE

In this book, the authors present current research in the study of fish ecology and marine ecosystems. Topics discussed include the metabolic rate of marine fish in early life and its relationship to their ecological status; antioxidant defense in marine fish; teleostean fish and the genomic content; the genetic structure of the masu salmon; the catadromous freshwater eel and the taxonomic representations of marine fish from archaeological assemblages in the Beagle Channel, Argentina.

Chapter 1 - Metabolic rate of different species varies as a consequences of both exogenous (temperature, salinity, oxygen consumption, pressure, feeding, pollution, etc.) and endogenous factors (animal size, growth, age, behavior, activity, physiological status and biochemical composition). Lipids area major energy source during fish early development, especially neutral lipids accumulated in yolk sac. Quantity of early life stages, larvae hatching and their survival depend on the energy generation, because energy is needed for synthesis of various substances, morphogenesis, successful growth and development. All these biological processes result in either heat generation or heat consumption. Direct microcalorimetry plays an important role for study these thermal activities because it is nondestructive and non-invasive to the animal. The goals of the present work were (i) to study the changes in lipid composition during the early life of some Black Sea fish species, (ii) to monitor the trends of heat production in embryos and hatching larvae using microcalorimetry and (iii) to evaluate the toxic effects of some organic substances – environmental pollutants on metabolic rate of fish early developmental stages. The heat dissipation of fish eggs and larvae was measured with a multi channel heat conduction type micro calorimeter (The Thermometric 2277 Thermal Activity Monitor, TAM, LKB, Sweden). Lipid composition fluctuations related with fish embryo genesis were shown. The level of triglycerides was decreased while the content of free fatty acids was elevated. Heat production of developing fish was increased during development and the high values were detected in hatching larvae. The obtained results demonstrated that metabolic rate correlated with the lipid content and reflected the peculiarities of fish early development stages depending of their ecology. The impact of different chemical stressors (pesticides, PCB, environmental pollution) modified metabolic rate of developing embryos and larvae associated with the disturbance of the processes of energy generation and utilization. The responses of fish early developmental stages on the complex of a biotic and anthropogenic factors associated with their ecological status are discussed.

Chapter 2 - The aerobic life is associated with the production of the potentially harmful components free radicals which damage many of biological molecules, membranes and cell components. So-called "oxygen paradox" means that the aerobic organisms including fish which cannot exist without oxygen are nevertheless inherently at risk due to oxidative stress. During the evolution process the defense mechanisms and complex of responses on oxidative stress were provided. They include special adaptive antioxidant enzymes and low molecular weight scavengers (vitamins, carotenoids, SH-compounds, and etc.). This complex protect organism against free radical production. At present increase of anthropogenic pollution of environment results negative biological effects in living organisms in all levels of their biological organization. Two consequences of environmental contamination might be noted: 1. xenobiotics cause the increase of oxyradical flux in the environment and result the deleterious in living organisms and 2. they induce the lipid peroxidation processes and the change of the prooxidant-antioxidant balance in the organisms which stimulate the pathologies, damage of immune system and premature senility and death.

Like all aerobic organisms, fish are very sensitive to oxidative stress and they have developed antioxidant defense. The study describes the various adaptation mechanisms and responses of marine fish (in the case of Black Sea especially) associated with oxidative stress. The results demonstrated that fish antioxidant status correlates with their phylogenetic position, trophic strategies, feeding behavior, variables of ecological conditions, swimming activity, belonging to the different ecological group (benthic, suprabenthic, suprabenthic-pelagic, and pelagic), seasonal fluctuations and some other environmental factors. The findings could be applied as for the solution of the problems of conservation ecology regarding pollution (especially in the case of Black Sea), but also for marine monitoring and fishery development.

Chapter 3 - The nature of the forces driving the genomic GC content of both prokaryotes and eukaryotes is matter of debate. Latest results favor selection as the main factor shaping base composition in bacteria. In vertebrates the subject is still under discussion.

Focusing on teleostean fish, the mass specific routine metabolic rate temperature-corrected using the Boltzmann's factor (MR) and base composition of genomes (GC%) were re-examined and related with their major habitats: polar, temperate, sub-tropical, tropical and deep-water. Fish of the polar habitat showed the highest MR. The MR of temperate fish was significantly higher than that of tropical one, which showed the lowest average value. Regarding GC%, polar and temperate fish both showed significantly higher values than both sub-tropical and tropical fish. Plotting MR and GC% a significant correlation was found between the two variables.

Different methylation levels characterized the genomes of fish living in different habitats. More precisely, the amount 5-methylcytosine (5mC) decreases from polar to tropical fish. Considering the positive correlation between CpG and GC%, as well as the temperature dependence of the 5mC deamination process, low genomic GC% in fish living in warm habitats could be the result of a CpG shortage.

The frequencies of CpG as well as that of the derivative doublets, that is TpA and CpA, were checked in the intronic sequences of the available teleostean genomes completely sequenced, namely *D. rerio*, *O. latipes*, *G. aculeatus*, *T. nigroviridis* and *T. rubripes.* The analysis of doublets further supported a link between environment, metabolic adaptation and genome base composition.

In this frame, also taking into account that MR turned out to be not significantly different between fish living in tropical and deep-water habitats, the authors suggest that the level of environmental O2 content (dictated by the Henry's law) could be a source of variability that directly affects MR adaptation and hence the base composition of fish genomes.

Neutral and selective hypotheses, proposed to explain the base composition evolution at the genome level, namely mutational bias, biased gene conversion (BGC), DNA breakpoints distribution, thermal stability and metabolic rate, were discussed in the light of current data. Negative (purifying) selection is proposed for the genome variation of teleostean fish among habitats.

Chapter 4 - Salmonid species have the potential to make a population genetic structure due to their homing behavior. However, the extent of genetic structuring would vary considerably because precision of homing varies within and among species. In addition to homing, several biotic and/or a biotic factors would also influence genetic structuring. A description of the genetic structure at various spatial scales and an understanding of the extent of genetic structuring could facilitate the identification of factors that affect genetic structuring. The knowledge of such factors is a fundamental requirement for the accurate inference of population dynamics, evolutionary processes, and conservation decisions. The authors focus on the factors that influence population genetic structuring and briefly describe the genetic structure of masu salmon (*Oncorhynchus masou*) populations at both regional and micro geographic scales. By analyzing mitochondrial DNA and microsatellite DNA variations, they found that masu salmon exhibit hierarchical genetic structuring and genetic differentiations not only at the regional scale but also at the microgeographic scale. These observations indicate that masu salmon would have the potential to make a population genetic structure at the microgeographic scale due to precise homing. Furthermore, it is also suggested that genetic structuring would be affected by several factors, such as refugia during glacial periods, ocean current, and dispersal patterns of each individual and the effects of such factors may vary depending on the intended geographic scale. However, such intrinsic genetic structuring faces the danger of being eroded by anthropogenic effects, including habitat degradation, habitat fragmentation, and artificial release of hatchery-reared fish. In fact, the negative impact of damming would hold true for masu salmon. The authors' results indicate the possibility that the indigenous genetic structure that has been created over many years would be lost by human activities during short periods.

Chapter 5 - The freshwater eel of the genus *Anguilla*, being catadromous, migrate between fresh water growth habitats and offshore spawning areas. However, a number of recent studies found that the temperate species Anguilla *anguilla*, *A. rostrata*, *A. japonica*, *A. australis* and *A. dieffenbachia* have never migrated into fresh water, spending their entire life in the ocean. Furthermore, those studies found an intermediate type between marine and freshwater residents, which appear to frequently move between different environments during their growth phase. The discovery of marine and brackish water residents suggests that anguillid eels do not all have to be catadromous and it calls into question the generalized classification of diadromous fish. However, there has been little available information concerning migration in tropical eels. In *A. marmorata,* showed three fluctuation patterns; (1) freshwater residence, (2) continuous residence in brackish water, and (3) residence in fresh water after recruitment, while returning to brackish water. Such migratory histories were found in other tropical eels, *A. bicolor bicolor* and *A. bicolor pacifica*. The *A. bicolor bicolor,* collected in a coastal lagoon of Indonesia, showed two patterns of habitat use, (1) constantly living in either brackish or sea waters with no fresh water life, and (2) habitat shift from fresh water to brackish or sea waters. The wide range of environmental habitat use indicated that migratory behavior of the tropical eel was facultative among fresh, brackish and marine waters during their growth phases after recruitment to the coastal areas. Further, the migratory behaviors of tropical eels appear to differ in each habitat in response to inter- and intra-specific competition. The results suggest that tropical eels have a flexible pattern of migration, with an ability to adapt to various habitats and salinities. This flexible habitat use was the same as that of temperate eels. Thus, the migrations of anguillid eels into fresh water is clearly not an obligatory behavior.

Chapter 6 - Native fish fauna of tropical semiarid freshwater ecosystems in Brazil are key representatives of the Neotropical region. Ichthyofauna of this region comprises of the orders Characiformes, Perciformes, Siluriformes and Synbranchiformes, besides invasive and exotic species. The population increase and the continued demand for quality fish generate the need for aquaculture as an alternative means of increasing fish production. Consequently, anthropic factors ranging from deforestation, construction of reservoirs on small rivers, introduction of invasive fish species to non-sustainable fish culture practice, can generate negative impacts on native species. Within these anthropic factors, proliferation of invasive and exotic fish species in the freshwater ecosystems has been deemed as a threat to the integrity of native species. However, the extent to which these factors can interfere and modify semiarid freshwater ecosystems remains an open question. This chapter focuses on how anthropic factors directly impact on the native fish species in semiarid freshwater ecosystems. Conversions of vegetation rich rural lands to urban areas degrade streams by qualitatively altering the composition, structure and function of their aquatic ecosystems. Construction of reservoirs can undoubtedly alter the natural hydrological regime of water bodies located downstream, and adversely impact on the migratory fish of this region. Introduction of exotic and carnivorous fish species from other hydrographical basins has significantly mediated to the decline of the native fish species, due to their competitive ability to thrive in semiarid aquatic ecosystems. Deforestation degrades riverine habitats and reduces ecological diversity of fish, besides causing pluvial shifts. Tropical semiarid Brazil is characterized with infrequent spells of rain interspersed with prolonged dry season. In tropics, rainfall is the main environmental driver that modulates the timing of fish spawning period. Changes in these environmental constraints present them with new challenges to survival and reproductive success. Since freshwater fish is important as a source of protein and provides sustained revenue, declining fish populations can have far reaching socioeconomic impacts which are compelling reasons for proper management. Strict regulations on introductions of fish species and restrictions on release of pollutants to aquatic ecosystems would reverse ecosystem degradation, improve water quality and conserve fish diversity.

Chapter 7 - This chapter analyses the taxonomic representations of marine fish from eight archaeological assemblages of the Beagle Channel. This study provides information about fish taxonomic distribution and ecological conditions in this region since 6400 to 500 radiocarbon years BP. Species corresponding to Nototheniidae family are represented throughout the archaeological sequence, in particular *Paranotothenia magellanica* and species attributable to *Patagonotothen* gender. Other fish of intertidal or shallow waters, as *Austrolycus depressiceps*, *Cottoperca gobio*, *Sprattus fueguensis*, etc., are also presented in the archaeological record. Among deep-sea species, *Macruronus magellanicus* is widely represented in the archaeofaunal assemblages. Nevertheless, the data also indicate that many species of deep waters (*Merluccius* sp. and *Thyrsites atun*)frequently consumed by hunter-gatherers are not common today in the marine ecosystem of the Beagle Channel. Comparisons of these data with modern ecological surveys indicate that both environmental stability at coastal waters and changes of the distributions of deep-sea species have occurred. This last result provides a new insight regarding the influence of modern human activities and the scope of overfishing in the uttermost part of the world.

Chapter 8 - Shads (Clupeidae: subfamily Alosinae) are a cosmopolitan group of fishes that exploit a wide range of habitats throughout the world, occurring in lakes, rivers, and seas. They are valuable ecological and economical resources. Species are able to adapt to estuarine, lentic and/or lotic habitats and present large plasticity in reproductive features, *e.g.* there is evidence of hybridization between allis shad *Alosa alosa* and twaite shad *A. fallax*. On the other hand, shads constitute important recreational and commercial fishes, *e.g.* American shad *A. sapidissima* in North America, allis shad in Europe or Indian shad *Tenualosa ilisha* in the Indo-Pak subcontinent and the Persian Gulf region. Most of these alosines are anadromous (a few landlocked populations have been found in Portugal, Ireland, Eastern Europeor Northern India), some species migrating several hundred kilometers upriver to spawn (*e.g.*800 km in the case of allis shad or as much as 1200 km in the case of Indian shad), and others exhibiting a pronounced homing behavior similar to that of migratory salmonids (*e.g.* allis shad). Adults are usually fished during the upriver movements towards spawning grounds but catches have been declining worldwide during the last 20 to 30 years, mostly due to anthropogenic activity, *e.g.* damming of rivers, overfishing (the predictability of migrations has rendered them vulnerable to overharvest) and deterioration of habitats by industrial and agricultural pollution.

Environmental conditions in the freshwater/brackish water reaches of rivers during the embryo-larval period are thought to play an important role in the future of populations, purposely determining the recruitment variability that is characteristic of the alosines. Moreover, the successful domestication of shads, American shad in USA and Reeves shad *T. reevesii* in China, and use as broodstock for restoration programs largely depends on understanding the biology of eggs, larvae and juveniles. Notwithstanding, the early life-history stages have been relatively poorly studied in European, Middle Eastern or African shads as opposed to North American species. A decade ago, two international conferences held in Bordeaux (France), in 1999, and in Baltimore (USA), in 2001, assembled the knowledge on world shads, namely their biology. Since then, relevant work on early life-history stages of shads worldwide has been carried out. Herein, the author compiles, updates and integrates these recent contributions on the biology and ecology of egg, larval and juvenile stages of shads, and prospects future work.

Chapter 9 - Heat shock proteins are highly homologous chaperone proteins, present in all cells playing key roles in limiting the consequences of protein damage and facilitating its recovery. There are accumulating data about the involvement of HSPs in chaperoning function. However, the task of HSPs with regard to its participation in cell survival processes still remains undeciphered in response to stress stimuli in natural conditions. The objective of this review is to interpret the role of HSPs in stress tolerance and cell survival mechanisms in fish exposed to exposed to pollutant stress in natural field conditions. HSPs inhibit apoptosis by down regulating apoptotic events like apoptosome formation, release of apoptogenic factors and ASK1-mediated JNK1/2 signaling etc. On the other hand, HSP also favors survival by promoting the activities of pro survival kinases like ERK1/2 and Akt. Generally, the overall balance between the signaling pathways of survival and death determine the fate of the cell. HSPs modulate multiple events within the signaling pathways in stressed fish thereby helping them to sustain survival.

In: Fish Ecology
Editor: Sean P. Dempsey

ISBN 978-1-61324-282-7

Chapter 1

METABOLIC RATE OF MARINE FISH IN EARLY LIFE AND ITS RELATIONSHIP TO THEIR ECOLOGICAL STATUS

Rudneva Irina, Shaida Valentin*
Institute of the Biology of the Southern Seas National Ukrainian Academy of Sciences, Nahimov av., 2, Sevastopol, 99011, Ukraine

ABSTRACT

Metabolic rate of different species varies as a consequences of both exogenous (temperature, salinity, oxygen consumption, pressure, feeding, pollution, etc.) and endogenous factors (animal size, growth, age, behavior, activity, physiological status and biochemical composition). Lipids are a major energy source during fish early development, especially neutral lipids accumulated in yolk sac. Quantity of early life stages, larvae hatching and their survival depend on the energy generation, because energy is needed for synthesis of various substances, morphogenesis, successful growth and development. All these biological processes result in either heat generation or heat consumption. Direct microcalorimetry plays an important role for study these thermal activities because it is nondestructive and non-invasive to the animal. The goals of the present work were (i) to study the changes in lipid composition during the early life of some Black Sea fish species, (ii) to monitor the trends of heat production in embryos and hatching larvae using microcalorimetry and (iii) to evaluate the toxic effects of some organic substances – environmental pollutants on metabolic rate of fish early developmental stages. The heat dissipation of fish eggs and larvae was measured with a multichannel heat conduction type microcalorimeter (The Thermometric 2277 Thermal Activity Monitor, TAM, LKB, Sweden). Lipid composition fluctuations related with fish embryogenesis were shown. The level of triglicerides was decreased while the content of free fatty acids was elevated. Heat production of developing fish was increased during development and the high values were detected in hatching larvae. The obtained results demonstrated that metabolic rate correlated with the lipid content and reflected the

* E-mail: svg-41@mail.ru

peculiarities of fish early development stages depending of their ecology. The impact of different chemical stressors (pesticides, PCB, environmental pollution) modified metabolic rate of developing embryos and larvae associated with the disturbance of the processes of energy generation and utilization. The responses of fish early developmental stages on the complex of abiotic and anthropogenic factors associated with their ecological status are discussed.

INTRODUCTION

Monitoring of fish development in early life is very important for the understanding of the mechanisms of gene expression and interaction between exogenous and endogenous factors. Undoubtedly the development of fish in early life depends on the specificity of their ecology and biology as well as on the adaptation to habitats. Studies of the metabolic strategies in embryogenesis of marine organisms are determined by various analytical techniques, including respirometry, biochemical and physiological analysis, direct calorimetry (Keckis & Schiemer, 1990; Canepa et al., 1997; Parra & Yufere, 2001). All biological processes result in either heat production or heat consumption and thus microcalorimetry is a versatile technique for a non-destructive and non-invasive method to animal to measuring total metabolism, independent of oxygen consumption (Baker & Mann, 1991; Russel et al., 2009). Besides that microcalorimetry is used for *in vivo* determinations and it's a great advantage as compared with the other biochemical and physiological methods.

Heat production of different species varies as a consequences of both exogenous (temperature, salinity, oxygen consumption, pressure, feeding, pollution, etc.) and endogenous factors (animal size, growth, age, behavior, activity, physiological status and biochemical composition) (Haegen et al., 1994; Lamprecht, 1998; Normat et al., 1998). The metabolic rate of the organism changes under the impact of many environmental factors including physical, chemical and biological (infection) and the toxic effects of them may be measured by multichannel microcalorimetry also.

Lipids are a major energy source during fish early development, especially neutral lipids accumulated in yolk sac (Mourente & Vazquer, 1996; Zhu et al., 1997).Quantity of early life stages, larvae hatching and their survival depend on the energy generation, because energy is needed for synthesis of various substances, morphogenesis, successful growth and development. The other problem of lipid metabolism in early stages is associated with the free radical production and lipid peroxidation processes that may cause pathology and diseases. For protection organism against toxic reactive oxygen species (ROS) impact antioxidant system is formed in early developmental stages in fish which includes specially adapted enzymes and low molecular weight antioxidants (Mourente et al., 1999a; Rudneva, 1999).

The effects of contamination on physiological and biochemical status of the organisms are well known and they led different kinds of the disturbances. Thus the measurements of respiration, locomotion, osmoregulation, pulse, biomarkers changes are provided for the evaluation of toxic effects of xenobiotics or environmental toxicity. In acute pollution and in high doses of pollutants exposure the physiological processes are damaged and often led death of the organism. In spite of this in many toxicological studies the survival (or mortality) of the test-organisms are used for toxicity evaluation and death as endpoint of the experiment is applied as insensitive bioindicator. Such physiological response gives little information about environmental pollution and its impact on the organism. Physiological and biochemical assays are useful for indicating the "early warning" biomarkers which induce at the beginning of stress reaction. Such a biomarkers affect in very low stressors concentrations and their fluctuations should be monitored and compared with the normal ranges. At present a lot of methods are used for biomarkers determinations. One of them is microcalorimetry which is possible to assay the metabolic rate in relation to activity organism and to estimate the energy costs at the normal and toxic conditions (Handy & Depledge, 1999).

Fish early developmental stages are very sensitive to environmental stressors both natural and anthropogenic. Xenobiotics accumulation, anoxia and hypoxia influence on metabolic rate (Penttinen & Kukkonen, 1998; Penttinen et al., 2005). Previously we described the polluted responses of fish eggs and larvae in the case of environmental pollution (Rudneva, 1998). Other studies indicate the changes of physiological and biochemical parameters during early life of marine organisms, which were associated with the changes of the metabolic rate (Canepa et al., 1996; Maenpaa et al., 2004; 2009; Penttinen et al., 2005). It's well-documented the relationships between pollutant exposure and metabolic rate of the test-organisms. But the trends of the heat production in toxicants-treated organisms are varied widely and in some cases the authors obtained contradictory results (Penttinen & Kukkonen, 2006). Taking into account that fish developing embryos and larvae are applied as test-organisms in evaluation of ecological status of marine areas it's important to obtain more detailed information of fish early life and its critical periods for application these data in aquaculture and for understanding the developmental mechanisms, metabolic strategy of the early developing stages, hatching process and the impact of environmental stress.

The goals of the present work were the following:

- to study the changes in lipid composition during the early development of some Black Sea fish species belonging to different ecological groups (benthic and pelagic);
- to monitor the heat production in embryos and hatching larvae using microcalorimetry method;
- to evaluate the toxic effects of some organic substances (environmental stressors) on the metabolic rate of fish developing embryos and larvae;
- to compare the metabolic rate of fish larvae collected in two locations with different level of pollution.

MATERIALS AND METHODS

Animals

Stage III eggs of Black Sea *Gobiidae Neogobius melanostomus (*Pallas*)* and *Proterorhinus marmaratus* (Pallas) were collected in the coastal waters of Sevastopol Bay and transferred into marine water (S=18‰) at the ambient temperature +20°C in the laboratory before hatching. Samples of eggs in developmental stages III-VI and hatching larvae were taken for lipid composition determination and VI stage eggs and hatching larvae were monitored for heat output measuring.

The fertilized eggs of Black Sea turbot *Psetta maxima maeotica* were obtained in the laboratory conditions, incubated at the ambient temperature and transported to 10-l incubation tanks in sterilized marine water (S=18‰) at + 16.6^{0}C before hatching. The embryos in various stages of development and new hatching larvae were used for lipid composition determination and metabolic rate assays.

1- day *Atherina hepsetus* and *Atherina mochon pontica* larvae (weight 3- 4 mg, length 8-8.2 mm) were collected in two sites of Sevastopol Bay differing of their ecological status, transferred in the laboratory and kept in aerated tanks with filtered sea water (10 individuals per 3 l) for further toxicological experiments and comparative studies of metabolic rate.

Experimental Set-up of Fish Development Studies

Lipid Composition Determination

The lipid composition was determined directly in the lipid extracts obtained by homogenizing fish eggs and larvae in hexane:isopropanol mixture (2:1, v/v) as we described previously (Rudneva, 1999). Lipid compounds were examined by thin-layer chromatography on Silufol plates (Kopytov, 1983). The plates were cleaned by chloroform and sprayed by phosphorus molibdenic acid in ethanol, then activated 5 min at +60° C. Lipid samples in hexane were brought on the plates in the concentrations 80-100 μg per plate. The chromatographic chamber was saturated by mixture hexane:ether (9:1 v/v). Silufol plates with lipid extracts were dipped into chloroform and separated twice. After separation the chromatograms were dried at +100°C. The ratio of lipid fractions were detected densitometrically used Beckmann Densitometer (Germany). All determinations were carried out in triplicates.

Metabolic Rate Determinations

The rate of heat production of fish VI stage eggs and hatching larvae was measured with a multichannel heat conduction type microcalorimeter (the Thermometric 2277 Thermal Activity Monitor, TAM). The TAM was fitted with four channels both containing a measuring and reference cell (twin calorimeter). The heat output signal was recorded by a Digitam Data program on PC and later analyzed by Digitam Data Analysis program (Thermometric, Sweden).

An individual *Gobiidae* egg or larvae was transferred and sealed alive in a 3-sm^3 glass measuring ampoule containing 2 ml of air-saturated sterile marine water. The reference ampoule contained 2 ml sterile marine water to avoid output cause by microorganisms. A baseline was measured both before and after the metabolic rate determination. Four individual measurements could be done simultaneously at the temperature of +20^0 C. Steady rates of heat dissipation were recorded within 1 h after the eggs were placed in the calorimeter cells. Heat output was continuously monitored over a period 130 h. The durations were identical for both *Gobiidae* species and determination of metabolic rate of eggs were measured before larvae hatching.

Two eggs of *P. maxima maeotica* were transferred in 2 ml of the sterile marine water and measured metabolic rate at +16.6^{0}C. Steady rates of heat dissipation were recorded within 1 h after the eggs were placed in the calorimeter cells and monitored due 80 h. Four individual measurements were done.

Simultaneously *Gobiidae* eggs were incubated in the glass cells at +20^{0}C and the developing embryos of *P. maxima maeotica* were incubated at +16.6^{0}C. The observations of the development and hatching processes of the non-stressed animals were carried out daily. The period of the embryogenesis and hatching time were identical in the intact and monitored animals.

Experimental Set-up of Toxicological Studies

Effect of PCB on Fish Eggs and Larvae

The effects of PCB (Arochlor 1254) were studied on developing embryos of *N. melanostomus* at the stage of IV and VI and hatching larvae. Each fish egg was sealed gently into 3-ml static glass microcalorimetric ampoule containing 2 ml of sterile air-equilibrated marine water (S=18‰) containing PCB in the concentrations of 10^{-4} and 10^{-3} μg l^{-1}. Control animals were incubated at the ampoule with 2 ml of sterile air-equilibrated marine water without toxicant. The reference ampoule contained 2 ml sterile marine water without sample. The differential heat output signal caused by sample was recoded with the program Digitam. The measurements were baseline – corrected using heat dissipation values of the empty ampoule both before and after monitoring.

The measurements started immediately and monitored directly in the TAM 2277 to receive information from the effects of rapidly response of metabolic rate during embryogenesis and hatching which was in the ampoule during the monitoring period. The metabolic rate was measured at the temperature of + 20^{0}C during 50 h according the protocol described above. All measurements were carried out in triplicates in both experimental and control groups of the animals

Because Arochlor was solved in the hexane the toxicity was studied in two groups of *A. hepsetus* larvae collected from the environment. In the first experimental group the individual larvae sealed gently into 3 sm^3 static glass ampoule contained air-equilibrated sterile marine water with hexane (Arochlor solvent) in the concentration of 1ml·l^{-1}. In the second experimental group the larvae was placed in the microcalorimetry ampoule contained sterile marine water and the mixture of hexane:Arochlor 1254 (10^{-3} μg·l^{-1}). The animals from the control group were incubated in sterile marine water without toxicant. The metabolic rate was measured at the temperature of +20^{0}C during 20 h according the protocol described above. All measurements in tested groups of animals were carried out in triplicates.

Effect of Pesticide Cuprocsat on Fish Larvae

A. mochon pontica larvae was sealed in the reference ampoule containing of 2 ml sterile air-equilibrated marine water with pesticide cuprocsat (CuSO4 $3Cu(OH)_2$ ·$1/2H_2O$) in the concentration of 0.156, 0.312, 0.625, 1.25 and 2.5 mg·l^{-1} The metabolic rate was monitored during 24 h at the temperature of +20^{0}C. Control animals were incubated at the ampoule containing 2 ml of sterile air-equilibrated marine water without pesticide. The reference ampoule contained 2 ml sterile marine water without sample. All determinations were carried out in triplicates.

Effect of Pesticide Cyfose on Fish Larvae

A. mochon pontica larvae was placed in the ampoule containing of 2 ml sterile air-equilibrated marine water with the pesticide cyfose in the concentration of 0.07 and 0.15 $mg l^{-1}$. The duration of the calorimetric measurements were identical as described above. All determinations were carried out in triplicates.

Statistical Analysis

Specific heat output was calculated as the average steady rate over the time of the monitoring. All numerical data are given as means $\pm$SD. The results were subjected to statistical evaluation with Student's *t*-test. Significant limits were set at $p \leq 0.05$. The correlation coefficients were calculated by the least-squares method between all examined lipid fractions content in every examined species and stage of the development. The comparison between dose of toxicants and heat output values were provided also using ANOVA and regression analysis (Halafian, 2008).

RESULTS

Lipid Composition Changes

The changes in lipid composition in developing embryos and hatching larvae of two fish *Gobiidae* species are presented in Figure 1. Dominating lipid components in early stage fish eggs were triglycerides which content was estimated as more than 50% of the total lipids. The interspecies differences in lipid composition were demonstrated also: alkoxylipids were identified in *N. melanostomus*, but not in *P.marmoratus*. Diglycerides level was higher ($p<0.05$) in *N. melanostomus* in all tested developmental stages as compared with *P. marmoratus*, while wax content showed the opposite trend.

Contents of phospholipides, mono- and diglycerides, fatty acid esters and wax varied insignificantly during embryogenesis and in larvae in both fish species. Sterols level increased during embryogenesis and in hatching larvae of *N. melanostomus* more than 2-fold as compared with the stage III egg ($p<0.05$). In *P. marmoratus* this parameter changed insignificantly. In both fish species free fatty acids considerable increased (in 2-4-fold) before hatching and in larvae as compared with the stage III eggs ($p<0.01$). Triglycerides level demonstrated the opposite trend: it dropped to the end of embryogenesis and in hatching larvae. Similar changes were detected for alkoxylipids content during *N. melanostomus* early life.

Thus the fluctuations of lipid composition during fish early development depended on the ecological and interspecies peculiarities. At the same time the average correlation coefficients between the content of lipid components were similar in the eggs and larvae of both fish species ($r=0.58$ for *P. marmoratus* and $r= 0.52$ for *N. melanostomus* respectively).

The ratios between triglycerides/phospholipids and sterol/phospholipids were estimated (Table 1). Factor triglycerides/phospholipids was declined from the stage III-IV eggs to the end of the embryogenesis and in hatching larvae. Factor sterols/phospholipids reflecting the cell membrane status was increased during early development in both fish embryos, but in *N. melanostomus* it was higher than in *P. marmoratus*. As we described above phospholipids content varied insignificantly during the examined period of development, but sterols concentration was grown. Thus sterols/phospholipids factor may reflect the modifications in cell membrane structure and changes of its fluidity during fish embryogenesis and in hatching larvae.

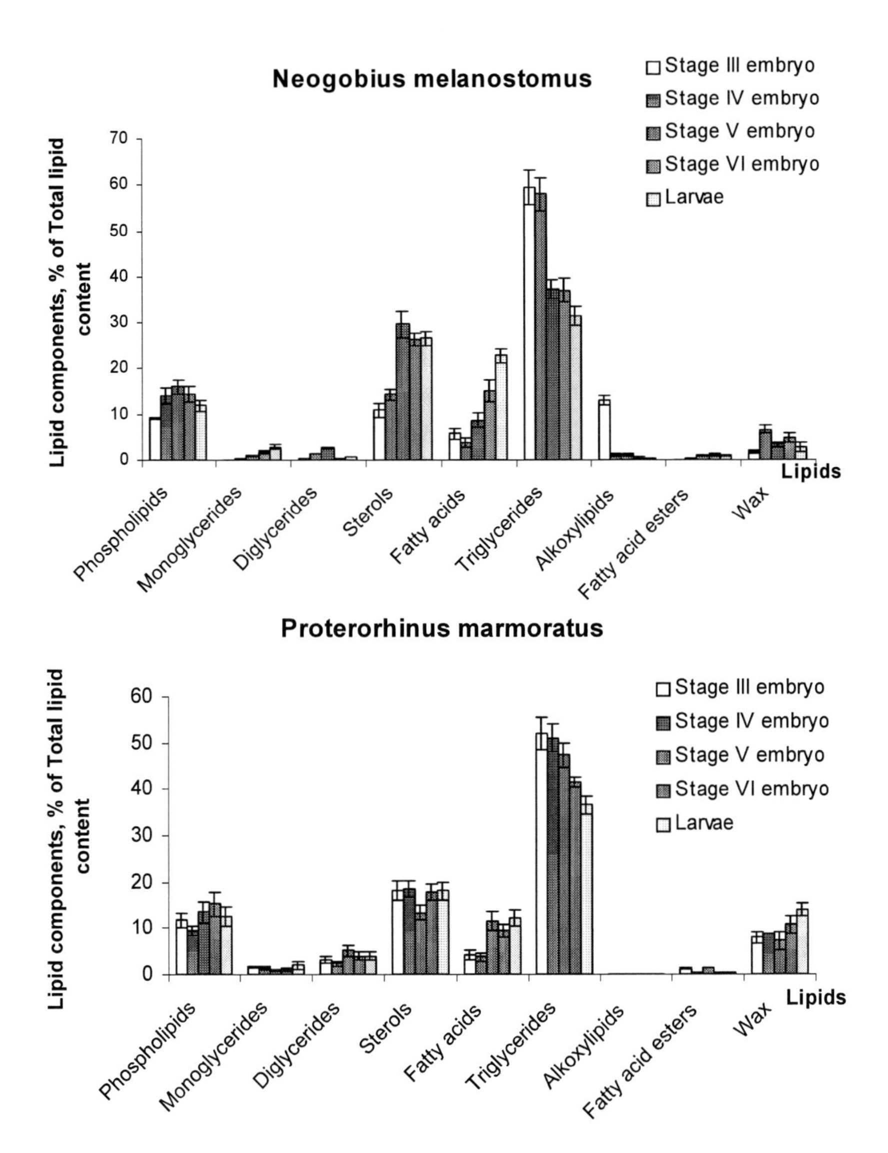

Figure 1. Lipid composition of developing eggs and hatching larvae of two Black Sea *Gobiidae* species. Average values are of triplicate determinations. Vertical bars indicate $\pm$SD.

The fluctuations in lipid composition in developing embryos in stages II-IV of Black Sea turbot *P. maxima maeotica* are presented in Figure 2.

The trends of lipid components concentrations in the embryos were not uniform. The content of phosholipids and fatty acids esters varied insignificantly in the examined developmental stages, concentrations of monoglycerides and fatty acids decreased to the stage V, diglycerides and sterols levels elevated at the stage III and dropped at the stage V. Wax concentration demonstrated the opposite trend. Level of triglycerides elevated during the examined period of development especially from stage II to stage III. The ratios between triglycerides/phospholipids and sterols/phospholipids are presented in Table 2.

Table 1. Ratio of major lipid components content in developing embryos and larvae of two Black Sea fish species

Stage of Develo-pment	Triglycerides/phospholipids		Sterols/phospholipids	
	N.melan-ostomus	*P.marm-oratus*	*N.melano-stomus*	*P.marmo-ratus*
III	6.49	4.45	1.17	0.70
IV	4.13	5.15	1.02	1.07
V	2.33	3.51	1.85	1.00
VI	2.59	2.71	1.83	1.15
Larvae	2.63	2.82	2.22	1.48

Factor triglycerides/phospholipids was elevated from the stage II-III embryo to the stage IV. The factor sterols/phospholipids was increased at the stage III and dropped at the stage IV. As we described previously phospholipids content changed insignificantly during the examined period, but the trends of sterols concentration was not uniform. Sterols/phospholipids ratio may reflect the modifications in cell membrane structure and its fluidity during fish embryogenesis. The average correlation coefficient between the examined lipids compounds was estimated as r=0.63 and it was higher than in *Gobiidae* embryos.

The obtained results demonstrated the differences of lipid composition fluctuations during early life of two species of *Gobiidae* and *P. maxima maeotica* (*Bothidae)* which should be associated with the specificity of their biology and ecology, because the *Gobiidae* eggs are benthic and *P. maxima maeotica* eggs are pelagic. Thus lipid composition and its trends during early development reflect closely the specificity of fish ecology and time of the early development.

Table 2. Ratio of major lipid components content in developing embryos of *P. maxima maeotica*

Stage of Development	Triglycerides/phospholipids	Sterols/phospholipids
II	2.09	0.85
III	2.49	1.17
IV	3.24	0.91

Metabolic Rate

Calorimetric tracings and mean rates of heat dissipation of developing stage VI embryos and hatching larvae of both Black Sea *Gobiidae* species are presented in Figure 3. The obtained results demonstrate the uniform trends. Hatching time was similar in both fish species. The metabolic rate increased constantly throughout the early development of both species and especially in hatching larvae. The reason of this fact lies in the active moving of the embryo for going out from the egg's shell.

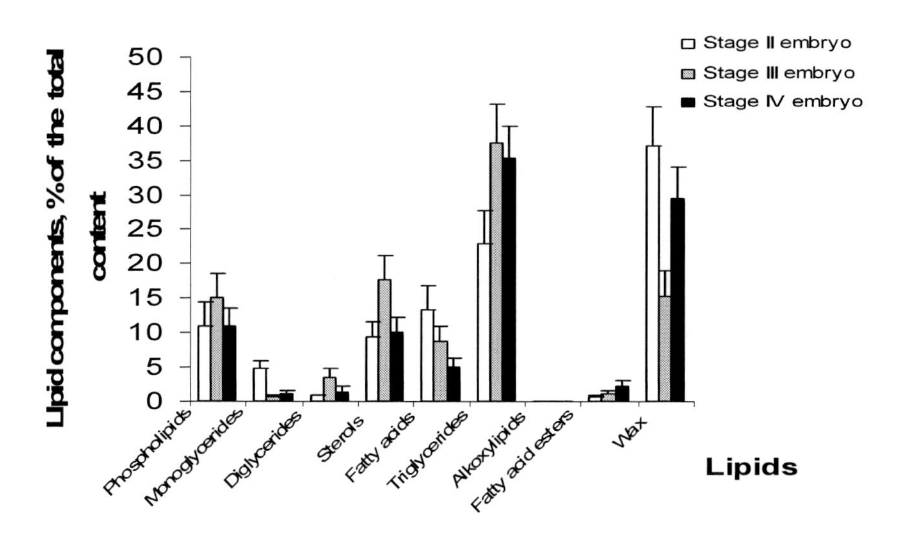

Figure 2. Lipid composition of developing eggs of *P. maxima maeotica*. Average values are of triplicate determinations. Vertical bars indicate ±SD.

Table 3. Statistical parameters of heat dissipation of two *Gobiidae* species developing eggs and larvae (n=4)

Parameters	*N. melanostomus*	*P. marmoratus*
Means of heat production, μW egg^{-1}, mean±SD	3.70± 0.36	3.20 ±0.52
Min/max heat production, μW egg^{-1}	5.90-30.11	4.90-18.98
Means of heat production, μW larvae^{-1}, mean±SD	14.28±1.32	12.12±1.22
Larvae heat production, % of the total, monitored during 130 h	91.93	87.75

The average rates of heat dissipation were similar in *N. melanostomus* and in *P. marmoratus* embryos (Table 3). At the same time the metabolic rate of individual egg of *N. melanostomus* varied between 5.9 and 30.1 μW egg^{-1} while in *P. marmoratus* the range was less (from 4.90 to 18.98 μW egg^{-1}) .

In spite of the differences between maximum value (more than 60%) of heat production in both fish species larvae the means were not significant. Larvae heat production monitored after hatching was also similar in both *Gobiidae* species.

Calorimetric tracings of three independent determinations of *P. maxima maeotica* eggs are presented in Figure 4. The curves are similar, the highest heat dissipation value was indicated in stage III of the development. The mean rates of heat dissipation (per mg of wet weight and per individual) of developing *P. maxima maeotica* embryos are presented in Figure 5. The hatching time of turbot at the temperature 16,6^{0}C was approximately 90 h.

As we see on the Figure 5 the metabolic rate of the individual egg varied between 0.9 μW and 2.2 μW, the highest values were detected at the stage III which was the longest period in the embryonic development. In the following developmental stages the heat dissipation varied insignificantly. The metabolic rate quota of each period of the embryonic development of turbot is presented in Figure 6. The obtained results demonstrate that more than 50% of the total heat dissipation associated with the stage III.

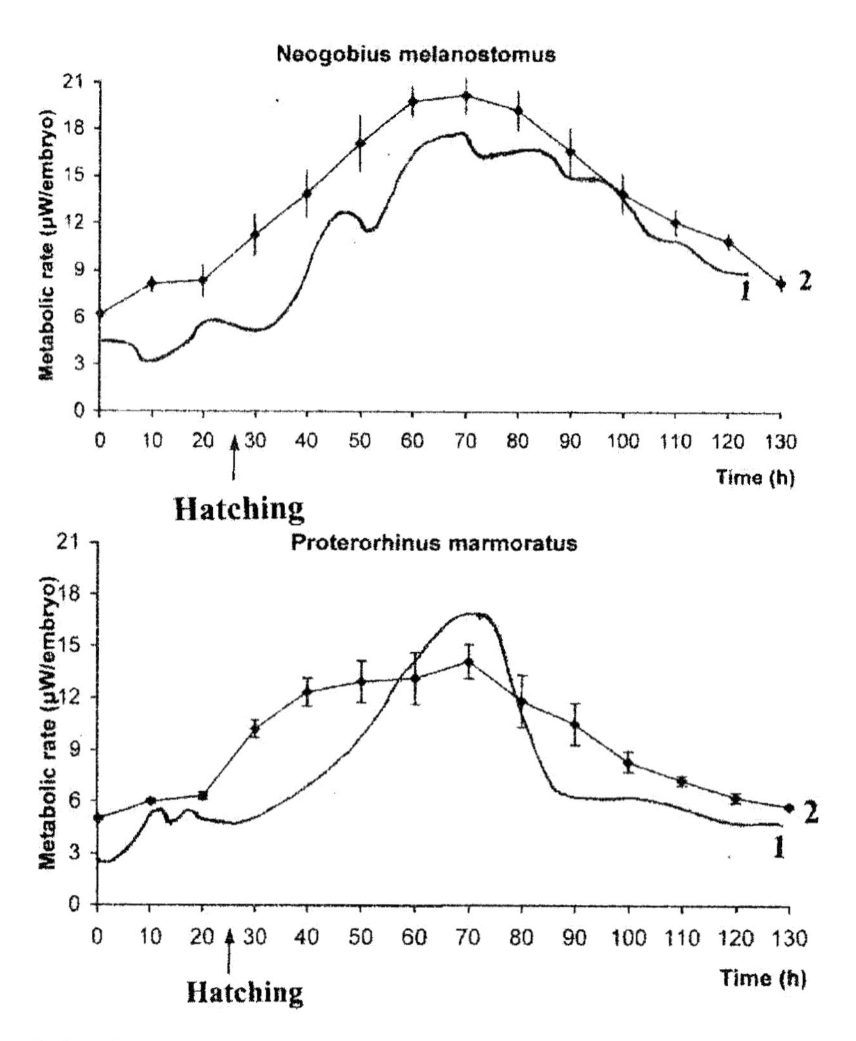

Figure 3. Calorimetric tracings (1) and mean rates of heat dissipation (2) of developing eggs and larvae of two Black Sea *Gobiidae*. Vertical bars indicate ±SD, n=4.

The obtained results demonstrated close link between the stages of the embryogenesis and the values of metabolic rate of the developing embryos and hatching larvae. We could conclude that the stage III was the longest and the important period of the turbot embryogenesis which characterized cell divisions, formation of the main organs, associated with the high metabolic activity and increase of energy consumption, respiration and oxidation. At the other periods of the embryonic development (stage IV-V) the embryo grows and the metabolic rate was relatively stable and then dropped to the end of the embryogenesis (stage VI).

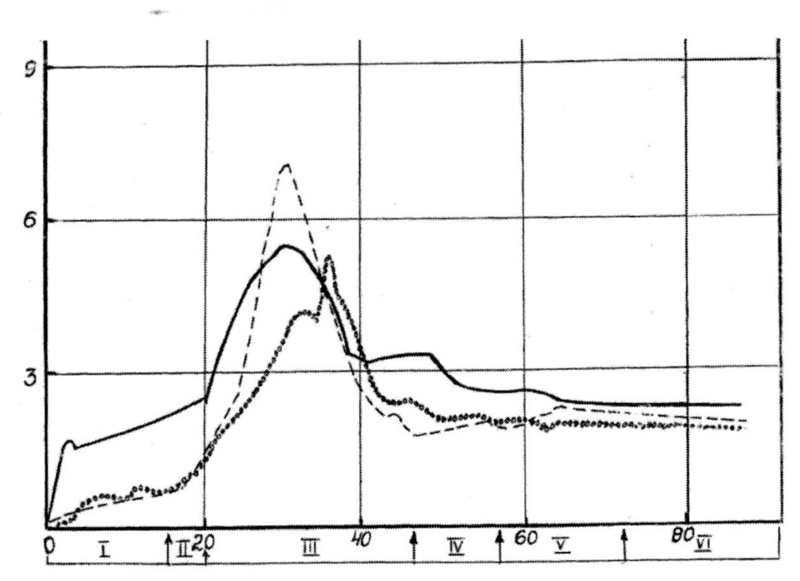

Figure 4. Calorimetric tracings of heat dissipation (three separate determinations) of developing embryos of *P. maxima maeotica*. The integral output signal was recorded once per 10 minutes, temperature of the incubation was +20^0 C, time of the monitoring was 80 h. Vertical axis means heat rate (μW) and horizontal axis means time of the monitoring (h). I – VI - stages of the embryo development.

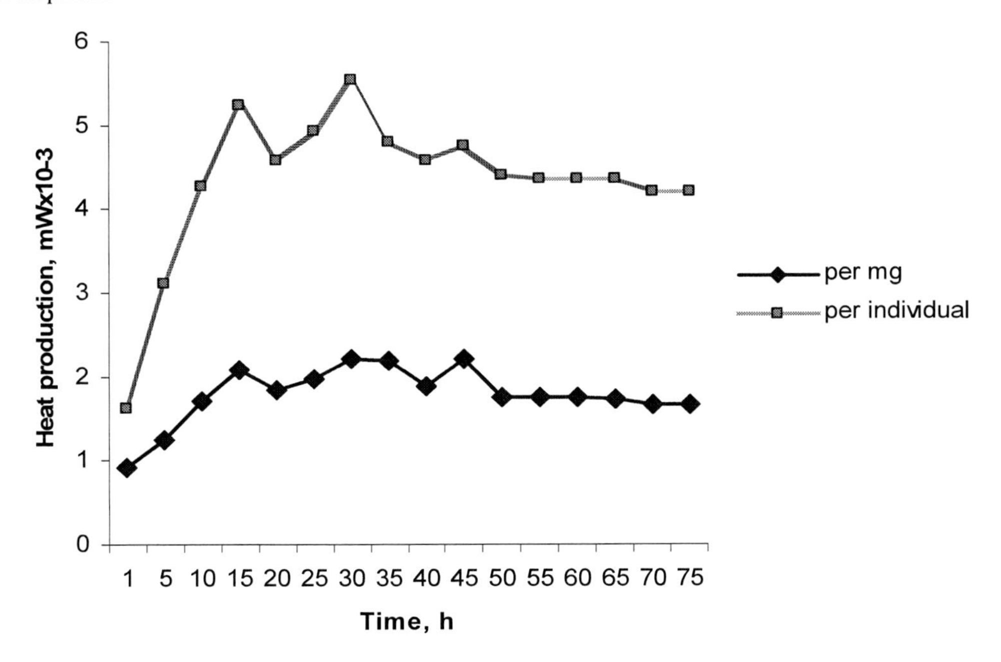

Figure 5. Calorimetric mean rates of heat dissipation of developing eggs and larvae of *P. maxima maeotica* (n=8). The integral output signal was recorded once per 10 minutes, temperature of the incubation was +16,6^0 C, time of the monitoring was 80 h.

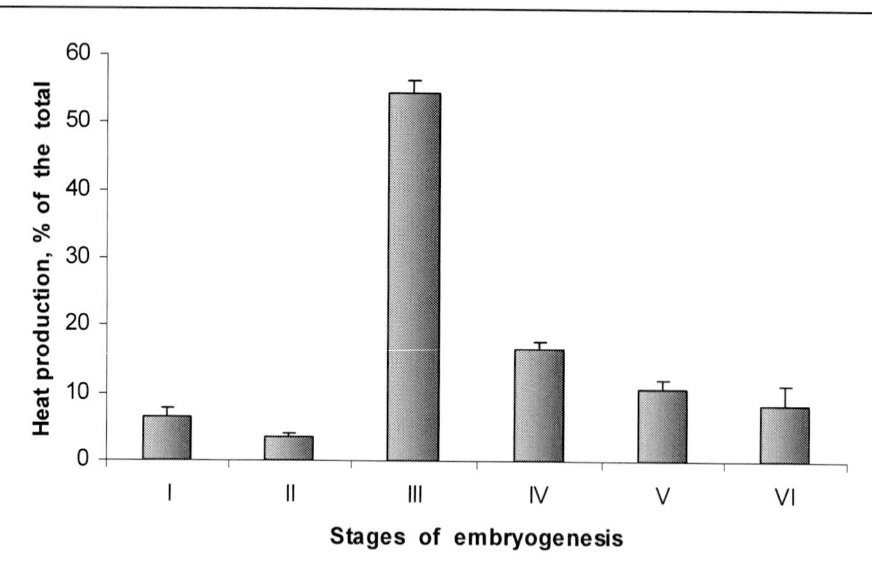

Figure 6. Heat production of developing eggs of *P. maxima maeotica* in different periods of embryogenesis. Temperature of the incubation was +16,6^0 C, time of the monitoring was 80 h.

Thus the present study demonstrates that metabolic rate in *Gobiidae* benthic eggs and larvae was similar while the heat production of pelagic developing embryos of *P. maxima maeotica* was differed significantly.

Effects of PCB on the Metabolic Rate of Developing Fish Embryos and Larvae

The polychlorinated biphenils (PCB), Cl-containing persistent organic compounds are highly distributed environmental pollutants due to their extensive application in different kinds of human activity. As other polyaromatic hydrocarbons (PAHs) they sequester in sediments in aquatic ecosystems and pose a hazard to the aquatic organisms especially benthic forms that are exposed to them directly or via food chains (Maenpaa et al., 2009). These xenobiotics impact marine organisms and cause damage of cellular and membrane components, biological molecules and result malformations of the developing eggs and embryos of fish and invertebrates. The aims of the present study were

- to test the toxic effects of PCB in benthic eggs of *N. melanostomus* and its hatching larvae using heat production values;
- to examine the toxicity of PCB solvent control (hexane only) in *N. melanostomus* eggs and larvae;
- to detect the toxicity effects in *A. hepsetus* larvae heat production;
- to compare toxicity effects in *N. melanostomus* larvae and *A. hepsetus* larvae

The effects of PCB in the concentrations of 10^{-4} and 10^{-3} $\mu g \cdot g^{-1}$ on the metabolic rate of developing eggs IV-VI stages and hatching larvae of round goby *N. melanostomus* were studied and the results are presented in Figure 7. No significant differences were shown in heat production in fish developing embryos in both examined stages, exposed in PCB in both concentrations during 50 h at the temperature +20^{0}C. However, heat production of the hatching larvae exposed in the water containing PCB in both tested concentrations was significantly less ($p<0.01$) as compared with intact animals.

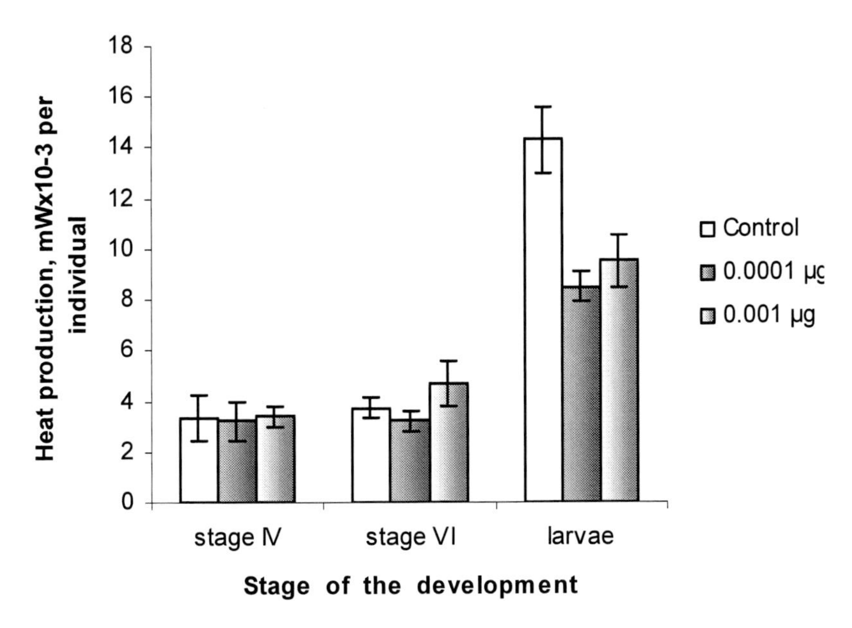

Figure 7. Metabolic rate of *N. melanostomus* developing embryos (stage IV and VI) and hatching larvae exposed in PCB in the concentrations of 10^{-3} and 10^{-4} µg l^{-1} (n=3, Mean$\pm$ SD). Time of the monitoring in the TAM was 50h at the temperature +20^0 C. The design of the experiment see above.

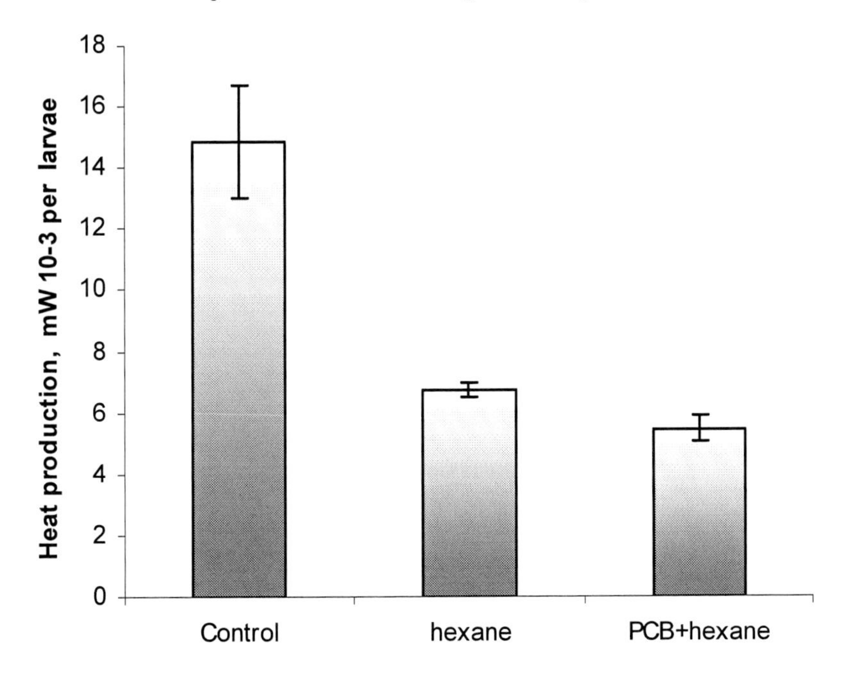

Figure 8. Metabolic rate of *A. hepsetus* larvae exposed in hexane (1 ml l^{-1}) and hexane +PCB (10^{-3}µg l^{-1}) (n=3, mean$\pm$ SD). Time of the monitoring in the TAM was 20h at the temperature +20^0 C. The design of the experiment see above.

Similar trends were shown in *A. hepsetus* larvae, incubated in solvent (hexane 1 ml l^{-1}) and in the mixture of hexane + PCB (10^{-3} μg l^{-1}) (Figure 8). Heat production of the animals exposed in both chemicals was significant less than the metabolic rate of the intact fish.

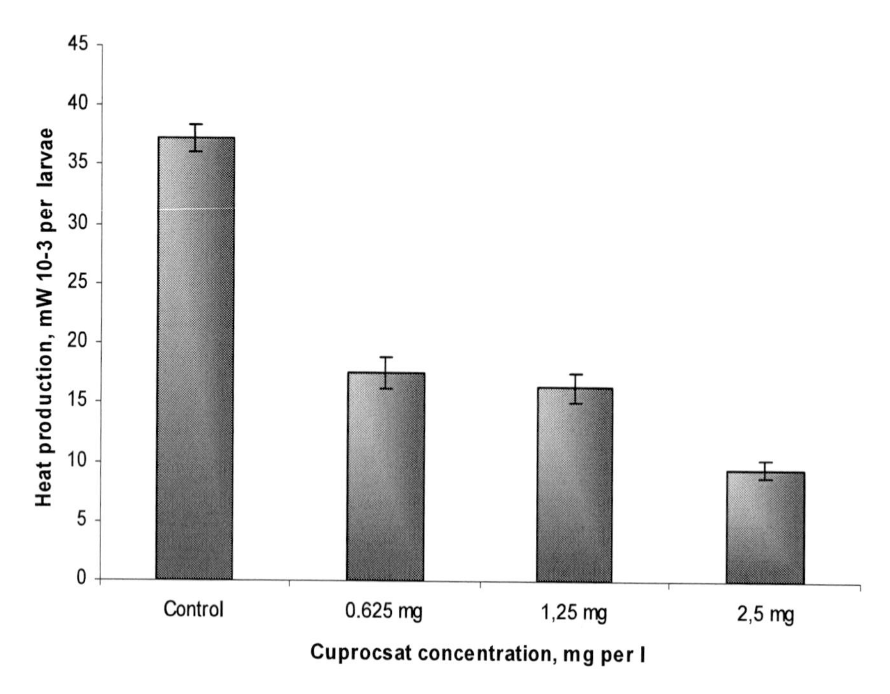

Figure 9. Metabolic rate *A. mochon pontica* larvae exposed in cuprocsat (n=3, mean± SD). Time of the monitoring in of the TAM was 24h at the temperature +20^0 C The design of the experiment see above.

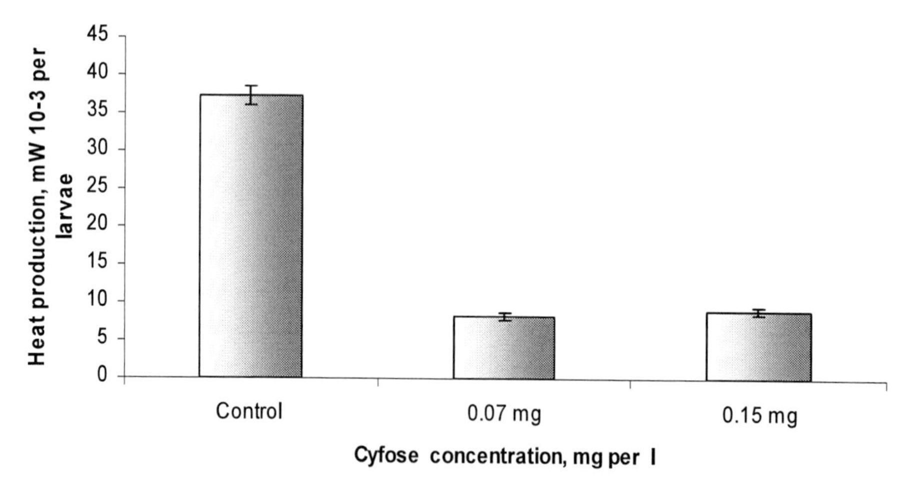

Figure 10. Metabolic rate *A. mochon pontica* larvae exposed in cyfose (n=3, Mean± SD). Time of the monitoring in of the TAM was 24h at the temperature +20^0 C The design of the experiment see above.

Thus, our finding demonstrated no significant differences between intact and PCB-treated fish eggs heat production in both tested concentrations of toxicant. At the same time fish larvae were more sensitive to toxicants and they showed significant decrease ($p<0.01$) of the metabolic rate in both tested PCB concentrations and solvent. *N. melanostomus* larvae heat production was dropped on 35-40% as compared with the intact animals while in treated *A.*

hepsetus larvae metabolic rate decreased on 54% in the case of solvent exposure and on 63% in the case of the mixture hexane+PCB. The obtained results show that larvae are more sensitive to PCB toxicity than eggs and *A. hepsetus* larvae heat production dropped more than *N. melanostomus* larvae. The general trend of metabolic rate decrease could be associated with the disturbance of energy generation and utilization processes and modification of the metabolic pathways.

Effects of Pesticides on the Metabolic Rate of Fish Larvae

Pesticides entering into marine ecosystem throw river discharge and domestic sewage led great stress for marine organisms, especially their early developmental stages. Cuprocsat ($CuSO4\ 3Cu(OH)_2\ 1/2H_2O$) and cyfose (phosphorus containing organic pesticide) are widely used in agriculture in the southern Ukraine region and study of their biological effects are very important for the development of the environmental monitoring and protection programs.

The goal of the present study was to examine the toxic response of fish larvae on both kinds of pesticides exposure. The results are presented in Figures 9 and 10.

Heat output of *A. mochon pontica* larvae exposed in cuprocsat in all tested concentrations was more than 2-fold less as compared with the value of the intact animals ($p<0.01$). At the same time the variables between the metabolic rate of the larvae exposed in the concentrations of 0.625 and 1.25 $mg\cdot l^{-1}$ were insignificant while the metabolic rate of the animals exposed in the highest concentration was significant less ($p<0.01$) than the intact and the other examined concentrations cuprocsat- treated fish.

Heat output of *A. mochon pontica* larvae exposed in both concentrations of cyfose decreased significantly ($p<0.01$) as compared with the control animals. However no differences were shown between the values of heat dissipation of both experimental groups.

Thus, we could proposed that cyfos was more toxic for fish larvae as compared with cuprocsat because significant decrease of heat production was indicated at less dose of organic pesticide as compared with Cu-containing chemical. High concentrations of cyfose were lethal for fish larvae and we could not measure their heat production in TAM.

Effects of Environmental Pollution on the Metabolic Rate of Fish Larvae

Environmental pollution includes many toxic xenobiotics (heavy metals, organic compounds, pesticides, PCB, DDT and their metabolites, etc.) which led very negative biological events both for marine ecosystem and aquatic organisms especially in early development stages. The goal of the present study was to evaluate the metabolic rate of *A. hepsetus* caught in highly polluted and non-polluted marine areas. The 1-day larvae (length 8.0-8.2 mm with small yolk sac) were collected in polluted area (port zone in Sebvastopol) and in non-polluted location near Chersones protection territory. The heat production trends of the larvae from both locations are presented in Figure 11.

The obtained results demonstrated the similar trends of heat production in both larvae during 50 h of monitoring in TAM. The curves reflect the period of stress and adaptation after the incubation in the TAM. At the same time at the period of normal physiological state (20-40h) heat output of the larvae from polluted site was significant less ($p<0.01$) than the values of the fish from non-polluted location. This fact could be explained the disturbance of the processes of energy generation and utilization in fish from the polluted habitats. However throughout the experiment the oxygen concentration and food in yolk sac have been decreasing and the differences between metabolic rate of two groups were less

Thus we could conclude that environmental pollutants are highly modifying factor of the metabolic rate of the early development stages of fish.

DISCUSSION

Description of Examined Fish Species Embryogenesis

Eggs of two examined fish species *N. melanostomus* and *P. marmoraus* are widely distributed fish in Black Sea, Azov Sea, Caspian Sea and Marmara Sea (Kalinina, 1976). They are differed each from other by some morphological parameters. The eggs of both species are benthic, but *P.marmoratus* has the smallest eggs among other *Gobiidae* species (height 2.5-2.6 mm and diameter 1.3-1.5 mm), while the eggs of *N. melanostomus* are greater (height 3.4-3.8 mm and diameter 1.7-1.9 mm). The embryonic development period is 10-14 days and it depends on water temperature which is ambient +20^0 C.

Embryogenesis of the *Gobiidae* species includes six developmental stages (Dechnick, 1973): stages I-II – eggs fertilization and blastula formation; stage III – gastrula formation and morphogenesis; stage IV – formation of the head, nervous system, cord, body segmentation; V – growth of the tail part, formation of the heart, intestine, liver, separation of the tail end from the yolk; VI stage – moving embryo, heart systole and respiration. The morphogenesis of the developing embryo requires high energy consumption and the major energy source are neutral lipids, accumulated in the yolk sac.

Opposite, Black Sea turbot *P. maxima maeotica* eggs are pelagic and their size is smaller than the *Gobiidae* eggs and their diameter varied between 1.1 and 1.83 mm (mean value 1.26 mm). Black Sea turbot eggs as another turbot species whose pelagic eggs contain a single oil globule. According the findings of T. Chesalina (Rudneva et al., 2001) embryogenesis of Black Sea turbot also includes six developmental stages and continues 4 days at the ambient temperature of +16.5^{0}C. The development begins (stage I) at 1.5-2 h after fertilization and continues 15 h. During stage II (5.5 h) the blastocel is formed, in the stage III (26 h) eyes, body segments, spinal cord and nerve system and pigmentation are presented in the embryo, in the stage IV (11 h) the further differentiation of the body continues, the intestine and heart system are appeared, in the stage V (15 h) the growth and development of the body are detected and in stage VI (18 h) the moving embryo is observed in the egg. At the end of the stage VI larvae hatches.

Egg size, ecological status and yolk sac composition closely associated with the period of embryonic development: in pelagic eggs it is shorter than in benthic ones. Fish eggs contain different energetic resources which involve into metabolism of the embryo in different phases of development and yolk dependence. Thus the changes of the main energetic substrates such as lipids may reflect the quality and quantity of developing processes in yolk-dependent fish embryos belonging to different ecological groups (benthic and pelagic). Lipids are the main components of yolk and the important source of the energy for developing embryo and their trends throughout embryogenesis are correlated closely with the total metabolic rate of the organism.

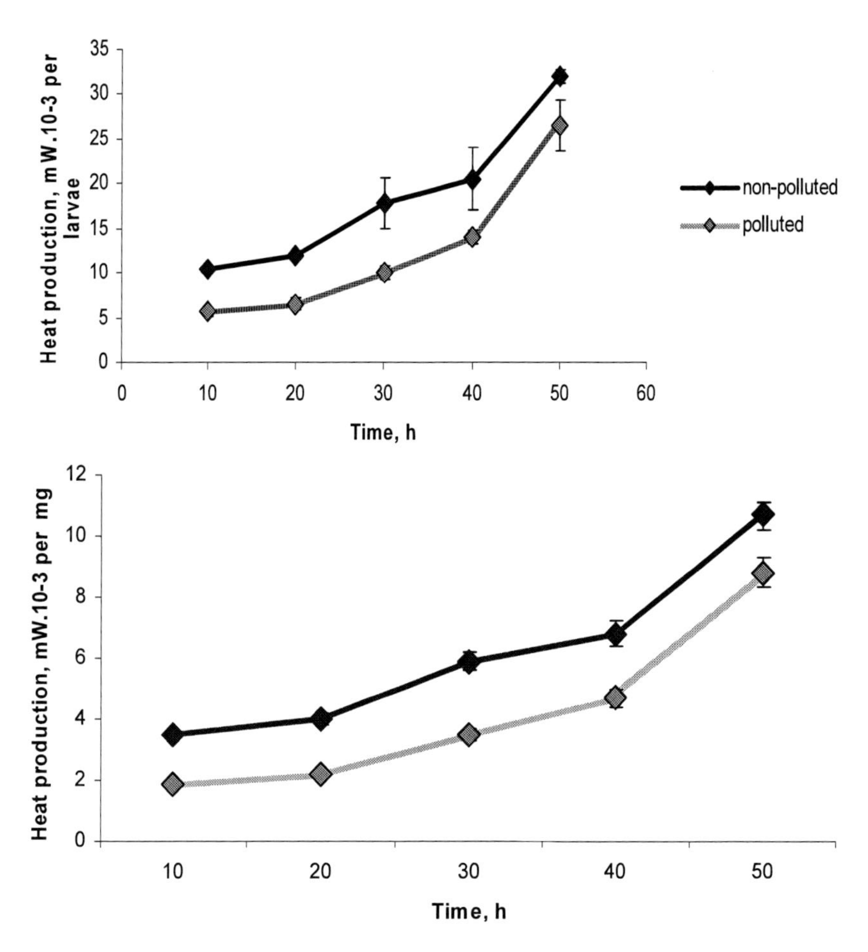

Figure 11. Metabolic rate of *A. hepsetus* larvae caught in polluted and non-polluted locations (n=5, mean± SD). Time of the monitoring in of the TAM was 50 h at the temperature +20^0 C The design of the experiment see above.

Lipid Composition during Fish Embryogenesis

To the end of the embryogenesis and in hatching larvae the concentration of triglycerides was declined and the level of free fatty acids was grown in both *Gobiidae* species. Triglycerides are the main energetic source in egg yolk and in some cases the success of larvae hatching and their further survival depend of these components concentration (Mourente et al., 1999a, b). During the early life the total lipid content decreased in the eggs of aquatic organisms, which directly correlated with the decline of triglycerides concentration. Lipid changes showed a permanent use of the endogenous reserve and no lipid synthesis was detected in any developmental stage (Roche-Meyzand et al., 1998). With this connection it was suggested that triglycerides may involve not only in energy generation, but they are used for phospholipids production, the main structural components of cell membranes in developing embryos. At the same time the changes in sterols/phospholipids ratio demonstrated the modifications in cell membrane fluidity in larvae as compared with the embryos which was documented by the other investigators also (Evans et al., 1997).

The interspecies peculiarities in lipid composition were found: in *N. melanostomus* alkoxylipid components were identified, but in *P. marmoratus* they were not shown. Additionally, lipid composition ratios were differed in two fish species eggs: content of diglycerides and wax were higher in *P marmoratus* eggs as compared with *N. melanostomus* while sterols and free fatty acids levels were less. Our data are agreed with the results of other investigators, who noted the specificity of yolk components in *N. melanostomus* (Moskalkova, 1984, 1985). According findings of Moskalkova, lipids were not determined in the embryo intestine and they were identified only in the yolk. The author suggested that the chemical composition of *N. melanostomus* embryo is specified which was connected with the peculiarities of the embryo feeding and food consumption. Circulation of yolk products and their metabolites was demonstrated during the embryogenesis of *N. melanostomus* and this process required additional energy (Moskalkova, 1985).

Throughout embryogenesis the eggs use different substrates for oxidation and energy generation. For instance, *Scophthalmus maximus* embryos demonstrated glycogen dependence following the first 18-19 h, free amino acids (84%) together with a small amount of phosphatidyl choline (9%) and later wax esters (5%) comprised the metabolic fuels of embryonic development. Following hatch (day 4.4 post fertilization) wax esters (33%) and triacylglycerols (25%) were initially catabolised with the remaining free amino acids (10%) (Finn et al., 1996). The results obtained in the present study show the fluctuations of lipid components level in developing embryos of *P. maxima maeotica*, but because we had not the opportunity to monitor the end of the embryogenesis and hatching larvae we could not discuss the general trends of the lipid composition changes. However, our findings demonstrated decrease of diglycerides and fatty acids level during embryo development from stage II to stage IV while the content of triglycerides varied insignificantly. Wax level dropped also from the stage II to stage IV. Our data agree with those cited above and we could conclude that neutral lipids and wax catabolised in the early phases of embryogenesis for energy generation and for cell division stage (Finn et al., 1996).

Decrease of lipid content especially neutral origin (triglycerides) is the general trend during early development of aquatic organisms and our data suggested this opinion. However, lipid composition and their ratio showed interspecies differences which depend of biological and ecological peculiarities of the examined species.

Metabolic Rate Trends

The rate of physiological processes in developing embryo is reflected in the metabolic rate and depends on the individual characteristics such as size, developmental phase, nutritional status and activity (Pakkasmaa et al., 2006). At the end of the embryogenesis the moving activity, respiration rate and heart systole were increased that also required high energy costs (Ozernuk, 1985). Throughout fish embryogenesis we demonstrated the variability of heat production rates, which could be connected with the involving of different energetic substances (carbohydrates, lipids) in the metabolism in different stages of the early development, dependence of yolk sac consumption, specificity of morphogenesis and growth

(Zhu et al., 1997; Desvilettes et al., 1997). These findings are suggested by the microcalorimetry studies which showed the growth of the heat production to the end of the embryogenesis in both examined *Goibiidae* species and especially in hatching larvae. The similar trends we noted in developing pelagic eggs and larvae of Black Sea turbot *P. maxima maeotica* at present study and in our previous publications (Rudneva & Shaida, 2000a; Rudneva et al., 2001). However, in pelagic turbot eggs which characterized short embryogenesis (4-5 days) the highest level of heat dissipation was detected in the stage III embryo (54,5% of the total heat production), at the period of intensive growth and morphogenesis. At the end of the embryogenesis metabolic rate in this species remained stable before hatching. Similar changes in heat production throughout fish early life of aquatic organisms were demonstrated by other investigators (Baker& Mann, 1991; Canepa et al., 1997; Geutner, 1998; Schmolz et al., 1999; McCollum et al., 2005; Pakkasmaa et al., 2006). Findings of heat production fluctuations in early life of terrestrial invertebrates were documented also (Penttinen & Holopainen, 1995; Lamprecht, 1998; Russel et al., 2009).

Table 4. Metabolic rate of some fish species in early development

Fish species	Developmental stage	Metabolic rate	Authors
Salvelinus alpinus	Eyed-stage egg	2.3-7.9 μW ind^{-1} 0.06-0.22 μW mg^{-1}	Pakkasmaa et al., 2006
	Hatching larvae	16.7 μW ind^{-1} 0.67 μW mg^{-1}	
Gadus morhua	Hatching larvae	2.14 μW ind^{-1}(unfed) 16.56 μW ind^{-1}(fed)	McCollum et al., 2006
	30dph larvae	12.6 μW ind^{-1}(unfed) 21.7 μW ind^{-1}(fed)	Geubtner, 1998
Neogobius melanostomus	Egg, stage V-VI Hatching larvae	3.70 μW ind^{-1} 14.30 μW ind^{-1}	Present study
Proterorchinus marmoratus	Egg, stage V-VI Hatching larvae	3.28 μW ind^{-1} 12.12 μW ind^{-1}	Present study
Psetta maxima maeotica	Egg, stage I-VI Hatching larvae	1.2-5.3 μW ind^{-1} 4.8 μW ind^{-1}	Present study
Atherina hepsetus	1-day larvae	5.2 μW ind^{-1}	Rudneva & Shaida, 2000a
Atherina mochon pontica	Larvae	36.6 μW ind^{-1}	Rudneva et al., 2004a; 2006a
Lepadogaster lepadogaster lepadogaster	Pre-larvae	1.63 μW ind^{-1}	Kuzminova, 2003

Besides that the metabolic rate of the individuals demonstrated high variability and closely depends both the environmental factors and genetic status also. For instance it was shown that the metabolic rates in Arctic charr *Salvelinus alpinus* eggs vary between families and eggs from different families can both genetically and maternally mediated (Pakkasmaa et al., 2006). Other investigators demonstrated increase of metabolic rate in feeding Atlantic cod *Gadus morhua* larvae as compared with unfed individuals (McCollum et al., 2005). Some data of metabolic rate during early development of the animals are presented in Table. 4.

At the same time resource availability may influence the costs of high metabolic rates which we demonstrated in eggs of fish species belonging to different ecological groups – benthic and pelagic. In both cases in spite of changes of embryogenesis time the observed increase of metabolic rate of developing fish embryos through early life is a general tendency in various fish species which probably caused by increase of oxygen consumption and respiration rate, heart systole, moving activity and stress of hatching, when embryo goes out to environment.

Thus, trends of heat production together with the changes of lipid composition could reflect the metabolic rate and correlation between energy generation and utilization in developing fish embryos and larvae. These results could be used for evaluation of critical periods of embryogenesis identification and hatching larvae survival which is important for aquaculture and ecotoxicologycal purposes. At the same time several metabolic processes affect by some chemical and physical factors which may change metabolic rate.

Chemical Stress and Metabolic Rate in Fish Embryo and Larvae

Environmental pollutants are highly modifying factor of the metabolic rate of the early development stages of fish. Metabolic costs are associated with maintenance/protection of the organism against environmental stressors (Penttinen et al., 2005; Penttinen & Kukkonen, 2006). At the same time metabolic rate can vary considerably with other causes than stressors. It tends to fluctuate due to many organism-specific factors such as growth, developmental stage, maturation, spawning, and etc. Thus it is very important to understand the mechanism of organism-specific variability in terms of energetic responses to environmental factors including pollutants.

At present study fish eggs and larvae were used as model systems to investigate the toxic effects of PCB and pesticides by the method of direct microcalorimetry as high sensitive tool required for measuring metabolic activity of the individuals by the monitoring of the heat output during early development and toxic exposure. Our findings demonstrated that heat production of early developmental stages of aquatic organisms (fish eggs and larvae) is highly sensitive to toxic effects and our results agree with those documented by other investigators (Maenpaa et al., 2004; Penttinen et al., 1996; 2006).

However, the results of metabolic rate modifications under the impact of different stressors are contradictory which reflects the difficulty of deriving and explaining of the ecotoxicological data (Table 5). The metabolic response should be separate through monitoring and the mechanisms of toxicity could be different and depended on the chemical structure and origin of xenobiotics and on tested organisms (Penttinen & Kukkonen, 1998; Penttinen et al., 2005).

Toxicity reduces metabolic rate which were demonstrated in moving activity, decrease of enzyme activities, blood vessel pulse in some invertebrates (Maenpaa et al., 2009). The ability to reduce metabolic rate during the exposure to environmental stress, termed metabolic rate suppression is an important component to enhanced survival in many organisms. Metabolic rate suppression can be achieved through modifications to behavior, physiology, and cellular biochemistry, all of which act to reduce whole organisms energy expenditure (Richards, 2010). In protozoa *Tetrahymena pyriformis* exposed in different kinds of toxicants the decrease of metabolic rate was documented (Beermann et al., 1999). Similar trends were determined in human blood cells treated by the chemicals dissolved in the water (McGuinnes & Barisas, 1991).

At the other hand some environmental stressors affect via a specific mechanism of toxicity such as uncoupling of oxidative phosphorilation, thus affecting the regulation of energy generation and dissipation and direct measurements the rate of energy metabolism demonstrated the increase metabolic rate of the animal impacted unfavorable factors (Widdows & Donkin, 1991). Besides that thyroid function may represent a sensitive target for environmental estrogenic compounds and the disturbance of endocrine metabolism may modify energetic metabolism also (Penttinen et al., 2005).

The results obtained on pentachlorphenol (PCP)-treated salmon eggs (0.992 $\mu mol \cdot l^{-1}$) demonstrated significantly increase the heat dissipation when compared with the ethanol (solvent) control and with intact group. In the higher concentrations the heat production decreased, which may indicate that PCP exposure was high enough to kill the fish embryo. PCP is a toxicant for uncoupling of phosphorilation and low concentrations of the PCP uncouples oxidative phosphorylation thus disrupting the regulation of energy metabolism by inhibiting the formation of ATP (Maenpaa et al., 2004). Penttinen & Kukkonen (2006) showed significant increase of heat production of the alevins of *Salmo salar* exposed in PCP in different concentrations. The output-mediated link between tissue residues concentration and heat output was demonstrated and the high correlation between bioaccumulation of PCP and exotoxicity was revealed. Larvae were more sensitive to the toxicants and embryo accumulated high amount of chemical in the late developmental stage possibly intensified the bioenergetic effect of PCP.

Thyroid- disrupting toxicants (bisphenol A and nonylphenol) were modified metabolic rate of salmon embryos throughout the development which was connected with the energetic consequences by affecting regulation of metabolic rate (Penttinen et al., 2005). Similar tendency was shown in fish larvae *Lepadogaster lepadogaster lepadogaster* (Bonnaterre) exposed in different concentrations of domestic sewage (1, 10 and 100 $ml \cdot l^{-1}$), containing the mixture of toxicants including the estrogens. The heat output of exposed larvae increased in series, but heat production of developing embryos of this fish species were not sensitive to the sewage impact (Kuzminova, 2003).

The modifying of heat dissipation under the impact of phenol chemicals (2,4-DNP, PCP, 2,4,5-TCB) were also demonstrated on aquatic invertebrates oligochaete *Lumbriculus variengatus*. The changes of heat output reflected the relationships between chemical concentrations and metabolic rate. Increase of metabolic rate was closely correlated with dose of toxicants (Penttinen & Kukkonen, 1998). At the same time the effect of toxicants on metabolic rate were differed in tested animals belonging to various taxonomic groups which demonstrate the different sensitivity of the animals to toxicants (Penttinen et al., 1996).

Table 5. Metabolic rate responses in fish species exposed in different toxicants

Fish species	Chemical substances	Metabolic response	Authors
Fish			
Eggs of *Salmo salar*	Pentachlorphenol (PCP)	Increase metabolic rate in low concentrations of PCP (0.992 $\mu mol\,l^{-1}$) and decrease in higher levels	Maenpaa et al., 2004
	Bisphenol A	Increase of metabolic rate	Penttinen et al., 2005
Alevins of *Salmo salar*	Pentachlorphenol	Increase of metabolic rate, high correlation between PCP accumulation and heat production	Penttinen & Kukkonen, 2006
Eggs and larvae of *Neogobius melanostomus*	PCB	Decrease of metabolic rate	Rudneva & Shaida, 2000b; present study
Atherina hepsetus larvae	PCB	Decrease of metabolic rate	Rudneva et al., 2000b; present study
Atherina mochon pontica larvae	Pesticide cuprocsat	Decrease of metabolic rate	Rudneva et al., 2004a; present study
	Pesticide cyfose	Decrease of metabolic rate	Rudneva et al., 2004b; present study
Atherina hepsetus larvae	Chronic environmental pollution		Rudneva et al., 1998, present study
Invertebrates			
Chironomid larvae *Chironomus riparius,* Oligochaete *Lumbriculus variengatus,* bivalves *Sphaerium comeum*	2,4,5-trichlorphenol (2,4,5-TCP)	Increase of metabolic rate almost 2-fold; No change Decrease of metabolic rate	Penttinen et al., 1996
Oligochaete *Lumbriculus variengatus*	2,4-dinitrophenol (2,4-DNP), PCP, (2,4,5-TCP)	Increase of metabolic rate was closely correlated with dose	Penttinen & Kukkonen, 1998
Nauplia *Artemia sp.*	PCB	Decrease of metabolic rate	Rudneva & Shaida, 2000 b
	Pesticide cuprocsat	Decrease of metabolic rate	Rudneva et al., 2004a; 2004c; 2005
	Pesticide cyfose	Decrease of metabolic rate	Rudneva et al., 2004b

Opposite, the results obtained at the present study showed that the heat production of examined fish larvae exposed in different concentrations of PCB was significantly dropped while the changes in eggs heat production were not shown. We could proposed that the tested PCB concentrations led high negative effects on fish larvae and disturbed the balance between heat generation and utilization. Previously we found the increase of reactive oxygen species production species and the changes in the antioxidant system responses in the tissues of adult fish species (Rudneva & Zherko, 1993; 1994; 1999; 2000). The similar result of metabolic rate decrease was noted in *Artemia* nauplia treated PCB (Rudneva & Shaida, 2000; Rudneva et al., 2006).

At the case of PCB-treated fish eggs and larvae we could note that larvae are more sensitive to the toxicant impact than the larvae. The similar effects were documented by the other authors also (Maenpaa et al., 2004; Penttinen & Kukkonen, 2006). In the case of fish egg the eggshell constitutes the main barrier for xenobiotics uptake into egg from water (von Westernhagen, 1998) and the larvae are more sensitive to chemicals impact. The uptake rate of chemicals was higher in the developmental stage approach in hatching which probably suggests increasing interaction of egg with surroundings environment and that eggshell supposed forfeits its role as a protective barrier along the development of the egg (Maenpaa et al., 2004). The strengthening of toxic effects in larvae may be related to the increasing intake of exogenous substrates from the environment also.

The comparative study of the pesticide effects on the metabolic rate demonstrated the similar decrease of the heat dissipation in fish larvae in spite of the different chemical structures of the tested pesticides. At the case of organic cyfose we could proposed that the mechanism of its toxicity and the influence on heat production was comparable with the effects of PCB and other xenobiotics of the similar origin (Rudneva et al., 2004b; 2006a,b).

As for cuprocsat the cooper containing pesticide the mechanism of its heat production modification in fish larvae could be differed. Metals are highly distributed pollutants in aquatic ecosystems and exposure to them disturbs metabolic processes in fish and invertebrates. The negative effects are characterized of changes in respiratory caused the damage of gill epithelium, inefficient oxygen delivery to tissues, changes in moving and energy requirements to resist or repair toxicant-induced damage (Hopkins et al., 2003; Pane et al., 2003).

It' well-documented that exposure to metals led the negative effects on aquatic animals including fish and invertebrates which associated with the damage of respiratory, swimming behavior, histopathology, accumulation in body tissues and damage of biochemical pathways. In our previous investigations we demonstrated the modification of protein and lipid metabolism, elevation the lipid peroxidation level and antioxidant enzyme activities in early developmental fish and invertebrates stages (Rudneva et al., 2005; Rudneva & Shaida, 2006; Zalevskaya et al., 2004). In many cases the investigators documented the increase of energy requirements to resist or repair toxicant-induced damage in aquatic animals. In contrast some metals are known to decrease metabolic rate or they don't change it (Hopkins et al., 2003). The reasons of the metabolism modification in aquatic animals exposed in metals could be explained with the hypometabolism likely stems from damage gill epithelium, neutoxicity or hypoactivity and these three factors that can also ultimately affect growth, survival and reproduction (Hopkins et al., 2003). At the same time despite of the high concentrations of Hg accumulated by mosquitofish in the experimental conditions no difference in metabolic rate among individual from three populations were found (Hopkins et al., 2003). In contrast the acute exposure to dissolved inorganic Hg increased metabolic rate of mosquitofish from the same source population (Tatara et al., 1999). The authors suggested that the chronic environmental contamination led the general resistance of the fish to Hg and subsequent attenuation of effects on metabolic rate (Hopkins et al., 2003). At our study cuprocsat cooper containing pesticide was toxic for fish larvae caused decrease of heat production. It should be consequences of metabolism reorganization and the enhance of reactive oxygen species (ROS) production which we demonstrated previously (Rudneva, 1998; Zalevskaya et al., 2004). ROS are damaged a lot of biological substances such as DNA, lipids and enzymes involving in the major metabolic pathways. High production of ROS impacted negatively on physiological status of the organism and modified its metabolism including energetic processes.

Chronic impact of environmental pollution on fish energetic metabolism was documented also (Hopkins et al., 2003; Pane et al., 2003; Rudneva et al., 1998). Our findings demonstrated that the fish larvae inhabited high polluted marine sites impact many toxicants which caused energetic metabolism disturbance (Hopkins et al., 2003). Thus many environmental factors biotic, abiotic and anthropogenic may modify fish metabolism and the application of microcalorimetry with other analytical methods should be helpful for the explanations of the mechanisms of the developmental changes during embryogenesis, the influence of xenobiotics disturbance and interactions between the developing embryo and environment.

Conclusions and Perspectives

1. Lipid composition and metabolic rate of developing fish embryos change throughout early life which caused the involving of different substrates for the processes of morphogenesis and for energy costs for these purposes. In all examined fish species belonging to different ecological groups (benthic and pelagic) metabolic rate increased during embryogenesis and especially in hatching larvae.
2. Heat production of examined fish larvae exposed in different concentrations of PCB and pesticides cuprocsat and cyfose was significantly decreased while the metabolic rate of the developing eggs exposed in tested toxicants were not changed. The toxic effect depends on the time of exposure and on the stage of the development.
3. Metabolic rate of *A. hepsetus* larvae inhabiting highly polluted area was significantly less as compared with the heat production of the larvae from the non-polluted area.
4. Heat output of aquatic organisms is very sensitive parameter for determination of water toxicity and thus it could be applied as biomarker to the evaluation of water quality and has taken an effective role in solving many environmental problems such as the interactions between environmental factors and fish development, toxic effects and fish protection. Sensitive endpoints should be used in risk assessment because already low environmental concentrations of toxicants may cause damage of the organism especially early developmental stages of fish.
5. Taking into account that at present the global climate changes modify the annual temperature and UV irradiation on the planet the microcalorimetry method could be applied for the evaluation of physical factors impact on aquatic organisms including

fish. Previously we demonstrated the post-effects on metabolic rate of crustacea *Artemia sp.* (Shaida & Rudneva, 2006) and fish could be very important test-organisms for study these effects.

ACKNOWLEDGMENT

We thank Dr. N. Shevchenko, Dr. T. Chesalina and Dr. N. Kuzminova for valuable assistance on the fish eggs collection, developmental stages and species determination and for participation in toxicological experiments.

REFERENCES

Baker S.M., Mann R.(1991). Metabolic rates of metamorphosis oysters (Crassostrea virginica) determined by microcalorimetry. *Am. Zool.* V. 31. P. 134 A.

Beermann K. H., Buchmann J., Scholmeyer E. (1999). A calorimetric method for the rapid evaluation of toxic substances using Tetrahymena pyriformis. *Thermochimica Acta.* V. 337 (1-2).P. 65 - 69.

Canepa E., Fraschetti S, Geraci S., Licciano M., Manganelli M., Alberyelli G., Riadi G. (1997). Microcalorimetry of some invertebrates: preliminary characterization of their metabolic activity during different developmental stages. *Biol Mar. Mediterr.* V.4. P. 626-628.

Dechnick T.V. (1973). *Ichthyophauna of Black Sea.* Kiev Naukova Dumka. 235 pp. (in Russian)

Desvilettes C., Bourdier G., Breton J.C. (1997). Changes in lipid class and fatty acid composition during development in pike (Esox lucius L) eggs and larvae. *Fish Physiol. Biochem.* V. 16. P. 381-393.

Evans R.P., Parrish C.C., Zhu P., Brown J.A., Davis P.J.(1997). Changes in phospholipase A (2) activity and lipid content during early development of Atlantic halibut (Hippoglossus hippoglossus). *Mar. Biol.* V. 130. P. 369-376.

Finn R.N., Fyhn H.J., Henderson R.J., Evjen M.S. (1996). The sequence of catabolic substrate oxidation and enthalpy balance of developing embryos and yolksac larvae of turbot (Scophthalmus maximus L). *Comp. Biochem. Physiol.* V. 115A (2). P. 133-151.

Geubtner J.A. (1998). Specific dynamics action, growth and development in larvae Atlantic cod Gadus morhua. *Thesis B.A.* University Delaware. 66 pp.

Haegan W.M. Vander, Owen R.B., Krohn W.B (1994). Metabolic rate of American woodcock. *Wilson Bulletin..* V. 106 (2). P. 338-443.

Halafian A.A. 2008. Statistica. Moscow, Binom Publ. 512 pp.

Handy R.D.& Depledge M.H. (1999). Physiological responses: their mesurements and use as environmental biomarkers in ecotoxicology. *Ecotoxicology.* V. 8. P. 329-349.

Hopkins W.A., Tatara C.P., Brant H.A., Jagoe C.H. (2003). Relationships between mercury body concentrations, standard metabolic rate and body mass in castem mosquitofish Gambusia holbrooki from three experimental pollution. *Environ. Toxicol. Chem.* V. 22. P. 586-590.

Kalinina E.M. (1976). *Reproduction and development of Azove Sea and Black Sea Gobiidae.* Naukova Dumka. Kiev. 118 pp. (in Russian)

Keckeis H., Schimer F. (1990). Consumption, growth and respiration of bleak, Alburnus alburnus (L.), and roach Rutilus rutilus (L.) during early ontogeny. *J. Fish Biol.* V.36.P. 841-851.

Kopytov J. P.(1983). New method of thin layer chromatography of lipids and carbohydrates. *Marine Ecology.* V.13. P 76-80 (in Russian).

Kuzminova N.S. (2003). The influence of different concentration of sewage on marine organisms. *Abstracts of the Int.Conf. "Space and Biosphere" Partenit,* Ukraine, Sept.28 – Oct. 4. Simpheropol. P. 164.

Lamprecht I.(1998).Monitoring metabolic activities of small animals by means of microcalorimetry. *Pure & Appl. Chem.* V. 70 (3). P. 695-700.

Maenpaa K.A., Penttinen O-P., Kukkonen J.V.K. (2004).Pentachlorophenol (PCP) bioaccumulation and effect on heat production on salmon eggs at different stages of development. *Aquatic Toxicology.* V. 68 (1). P. 75-85.

Maenpaa K.A., LeppanenM.T., Kukkonen J.V.K.(2009). Sublethal toxicity and biotransformation of pyren in Lumbriculus variegates (Oligochaeta). *Science of the Total Environment.* V. 407. P. 2666-2672.

McCollum A., Geubtner J., von Herbing H. (2005). Metabolic cost of feeding in Atlantic Cod (Gadus morhua) larvae using microcalorimetry. *ICESJ. Marine Science.* V. 63 (3). P. 333- 339.

McGuinnes S.M., Barisas B.G. (1991). Acute toxicity measurements on aquatic pollutants using microcalorimetry on tissue-cultured cells. *Environ. Sci.Technol.* V. 25 (6). P. 1092 - 1098.

Moskalkova K.I. (1984). Unusual way of embtyonic feeding in Neogobius melanostomus (Pallas) (Pisces, Gobiidae). *Reports of the USSR Academy of Sciences 278*, 1127- 1130 (in Russian)

Moskalkova K.I. (1985). "Couprofagia" in embryos of teleost fish Neogobius melanostomus (Pallas) (Pisces, Gobiidae). *Reports of the USSR Academy of Sciences.* V.282.P. 1251-1254.

Mourente G., Vazquez R. (1996). Changes in the content of total lipid, lipid classes and fatty acids of developing eggs and unfed larvae of the Senegal sole, Solea senegalensis Kaup. *Fish. Physiol. Biochem.* V. 15. P. 221-235.

Mourente G., Yocher D.R., Diaz E., Grau A., Pastor E. (1999a). Relationships between antioxidants, antioxidant enzyme activities and lipid peroxidation products during early development in Dentex dentex eggs and larvae. *Aquaculture.*V. 179. P. 309-324.

Mourente G., Rodriguez A., Grau A., Pastor E. (1999b). Utilization of lipids by Dentex dentex L. (Osteichyhyes, Sparidae) larvae during lecitotrophia and subsequent starvation. *Fish Physiol. Biochem.* V. 21. P. 45-58.

Normat M., Graf G., Szaniawska A. (1998). Heat production in Saduria entomon (Isopoda) from Gulf of Gdansk during an experimental exposed to anoxia conditions. *Marine Biol.* V. 131. P. 269-273.

Ozernuk N.D. (1985). Energetic metabolism of fish in early ontogenesis. *Nauka.* Moscow. 175pp. (in Russian)

Pakkasmaa S., Penttinen O-P., Piironen J. (2006). Metabolic rate of Asrctic charr eggs depend on their parentage. *J. Comp. Physiol.*V.176B. P. 387-391.

Pane E.F., Smith C., McGreer J.C., Wood C.M. (2003). Mechanism of acute and chronic waterborne nickel toxicity in the freshwater cladoceran Daphnia magna. *Environ. Dci. Technol.* V. 37. P. 4382-4389.

Parra G. & Yufera M. (2001). Comparative energetics during early development of two marine fish species. Solea senegalensis (Kaup) and Sparus aurata (L.). *J. Exp. Biol.* V. 204. P. 2175-2183.

Penttinen O-P.& Holopainen J. (1995).Physiological energetics of a midge, Chironomus riparius Meigen (Insecta, Diptera): normoxic heat output over the whole life cycle and response of larva to hypoxia and anoxia. *Oecologia.* V. 103. P. 419-424.

Penttinen O.P., Kukkonen J., Pellinen J. (1996). Preliminary study to compare residues and sublethal energetic responses in benthic invertebrates exposed to sediment-bound 2,4,5-trichlorophenol. *Environ. Toxicol. Chem.* V. 15. P.160-166.

Penttinnen O.P., Kukkonen J.V.K . (1998). Chemical stress and metabolic rate in aquatic invertebrates: threshold, dose-response and mode of toxic action. *Environ. Toxicol. Chem.* V. 17. P. 883-890.

Penttinnen O.P., Honkanen J.O., Sorsa K., Kukkonen J.V.K.(2005).Can aquatic pollutants cause specific endocrinological and metabolic responses in salmon (Salmo salar m. Sebago) embryos? A direct calorimetry study. *Verhandlungen International Vereinigung Limnology.* V. 29. P. 945-948.

Penttinen O-P.& Kukkonen J.V.K. (2006). Body residues as dose for sublethal responses in alevins of landlocked salmon (Salmo salar m Sebago): a direct calorimetry study. *Environmental Toxicology and Chemistry.* V. 25.(4). P. 1088-1093.

Richards J.G. (2010). Metabolic rate suppression as a mechanism for surviving environmentalchallenge in fish. In: *Progress in Molecular and Subcellular Biology.* V. 49. P. 113-135.

Roche-Mayzaud O., Mayzaud P., Audet C. (1998). Changes in lipid classes and tripsin activity during the early development of brook charr, Salvelinus fontinalis (Mitchill), fry. Aquacult. Res. V. 29. P. 137-152.

Rudneva I.I. (1998). The biochemical effects of toxicants in developing eggs and larvae of Black Sea fish species. *Marine Environ. Res.* V. 46. № 1-5. P. 499-500.

Rudneva I.I. (1999). Antioxidant system of Black Sea animals during early development *Comp. Biochem and Physiol.* V. 122C. p. 265-271.

Rudneva I.I. & Zherko N.V. (1993). The effect of PCB on lipid and protein metabolism of Black Sea scorpion fish. *Reports of the Academy of Sciences of Ukraine.* V.11. P. 157-161 (in Russian)

Rudneva I.I. & Zherko N.V. (1994). The effect of PCB on antioxidant enzyme activities and lipid peroxidation in muscle and liver tissues of two Black Sea fish species. *Biochemistry.* V. 59 (1). P.. 34-44 (in Russian)

Rudneva I.I., Zherko N.V. (1999). The effect of PCB on antioxidant system and lipid peroxidation in gonads of Black Sea red mullet Mullus barbatus ponticus. *Marine Biology* V. 25. P. 239-242. (in Russian)

Rudneva I.I., Zherko N.V. (2000).The effect of PCB on antioxidant system and lipid peroxidation in gonads of Black Sea Scorpion fish. *Ecology.* V. 1. P. 70-73. (in Russian)

Rudneva I.I., Chesalina T.L., Shaida V.G., Shevchenko N.F(1998). Morphology and heat production in atherina larvae (Atherina hepsetus L.) from contaminated and non-contaminated regions. *Marine Ecology.* V. 47. p. 33-36. (in Russian)

Rudneva I.I. & Shaida V.G. (2000a). Microcalorimetric studies of marine organisms early development. *Microcalorimetric Studies in Marine Biology. ECOSY. Sevastopol. P.* 139-149. (in Russian)

Rudneva I.I. & Shaida V.G. (2000b). Microcalorimetric studies of PCB effects on early developmental stages of marine organisms. *Microcalorimetric Studies in Marine Biology. ECOSY. Sevastopol. P.* 149-157. (in Russian)

Rudneva I.I., Chesalina T.L., Shaida V.G. (2001). Physiological and biochemical characteristics of Black Sea flounder Psetta maxima maeotica embryogenesis. *Ichthyology.* V. 41.P. 717-720 (in Russian).

Rudneva I.I., Shaida V.G., Kuz'minova N.S. (2004a). The fungicide cuprocsat effect on the Artemia and Atherina larvae heat production. *Agroecological J.* No 3. P. 81-83.(in Russian)

Rudneva I.I., Shaida V.G., Kuz'minova N.S., Kutsuruba I.E. (2004b). Analysis of the cyfose toxicity with the use of Artemia salina nauplia. *Agroecological J.* No 3. P. 57-62.(in Russian)

Rudneva I.I.,Zalevskaya I.N., Kuz'minova N.S. Savkina Y.G. (2004c). Evaluation of fungicide cuprocsat tpoxicity on the Black Sea Atherina larvae. *Agroecologica*l J.. No 3. P. 83-89. (in Russian)

Rudneva I.I., Shaida V.G., Kuz'minova N,S., Kuzuruba I.E. (2005). Artemia salina L. as test-organism in the evaluation of the toxicity of fungicide cuprocsat. *Biology of the indoor waters.* No 3. P. 104 – 109.

Rudneva I.I.& Shaida V.G. (2006). Metabolic strategy in the early life of some Black Sea fidh species measured by microcalorimetry method. *XIVth Conference The Amber ISBC. Abstr. Sopot,* Poland, June 2- 6, 2006. P.74.

Rudneva I., Shaida V., Kuzminova N.S. (2006). Application of microcalorimetry for toxic effects evaluation in aquatic animals. *XIVth Conference The Amber ISBC. Abstr. Sopot,* Poland, June 2- 6, 2006. P.77

Russel M., Yao J., Chen H., Wang F., Zhou Y., Choi M.M., Zaray G., Trebse P. (2009). Different technique of microcalorimetry and their application to environmental sciences: a review. *J. American Science.* V. 5(4). P. 194-208.

Schmolz E., Drutschmann S., Schricker B., Lamprecht L.(1999). Calorimetric measurements of energy contents ad heat production rates during development of the wax moth Galleria mellonella. *Thermochimica Acta.* V. 337. P. 83-88.

Shaida V.G.& Rudneva I.I. (2006). The influence of climate modulating factors on the heat production of Crustacea measuring microcalorimetry method. *XIVth Conference The Amber ISBC. Abstr. Sopot,* Poland, June 2- 6, 2006. P.129.

Tatara C.P., Mulvey M., Newman M.C. (1999). Genetic and demographic responses of mosquitofish (Gambusia holbrooki) exposed to mercury fro multiple generations. *Environ. Toxicol Chem.*V. 18. P. 2840-2845.

von Westernhagen H. (1988). Sublethal effects of pollutants on fish eggs and larvae. In: *Fish Physiology. V. XI. The Physiology of Developing Fish. Par A Eggs and Larvae.* (Hoar W.S., Randall D.J. Eds). Academic Press Inc. San Diego. P. 253-346.

Widdows J. & Donkin P. (1991). Role of physiological energetics in ecotoxicology. *Comp. Biochem Physiol.* V. 100. C. P. 69-75.

Zalevskaya I.N., Matveeva Z.S.,Rudneva I.I. (2004). Evaluation of the fungicide cuprocsat toxicity on Artemia salina *Agroecological J.* No 3. P. 75-78. (in Russian)

Zhu P., Evans R.P., Parrish C.C., Brown J.A., Davis P.J. (1997). Is there a direct connection between amino acid and lipid metabolism in marine fish embryos and larvae? *Bull. Aquacult. Assoc. Can.* V. 97-2.P. 48-50.

In: Fish Ecology
Editor: Sean P. Dempsey
ISBN 978-1-61324-282-7

Chapter 2

ANTIOXIDANT DEFENSE IN MARINE FISH AND ITS RELATIONSHIP TO THEIR ECOLOGICAL STATUS

Irina Rudneva*
Institute of the Biology of the Southern Seas National Ukrainian Academy of Sciences, Nahimov av., 2, Sevastopol, 99011, Ukraine

ABSTRACT

The aerobic life is associated with the production of the potentially harmful components free radicals which damage many of biological molecules, membranes and cell components. So-called "oxygen paradox" means that the aerobic organisms including fish which cannot exist without oxygen are nevertheless inherently at risk due to oxidative stress. During the evolution process the defense mechanisms and complex of responses on oxidative stress were provided. They include special adaptive antioxidant enzymes and low molecular weight scavengers (vitamins, carotenoids, SH-compounds, and etc.). This complex protect organism against free radical production. At present increase of anthropogenic pollution of environment results negative biological effects in living organisms in all levels of their biological organization. Two consequences of environmental contamination might be noted: 1. xenobiotics cause the increase of oxyradical flux in the environment and result the deleterious in living organisms and 2. they induce the lipid peroxidation processes and the change of the prooxidant-antioxidant balance in the organisms which stimulate the pathologies, damage of immune system and premature senility and death.

Like all aerobic organisms, fish are very sensitive to oxidative stress and they have developed antioxidant defense. The study describes the various adaptation mechanisms and responses of marine fish (in the case of Black Sea especially) associated with oxidative stress. The results demonstrated that fish antioxidant status correlates with their phylogenetic position, trophic strategies, feeding behavior, variables of ecological conditions, swimming activity, belonging to the different ecological group (benthic, suprabenthic, suprabenthic-pelagic, and pelagic), seasonal fluctuations and some other environmental factors. The findings could be applied as for the solution of the problems

* E-mail: svg-41@mail.ru

of conservation ecology regarding pollution (especially in the case of Black Sea), but also for marine monitoring and fishery development.

1. INTRODUCTION

Aerobic life is associated with oxygen consumption and involving it into the major metabolic pathways of the living organisms. At the other hand O_2 has two unpaired electrons and the univalent reduction of the molecular oxygen produces reactive oxygen species (ROS) such as superoxide radical (O_2^-), singlet oxygen ($^1O_2$), hydrogen peroxide (H_2O_2), hydroxyl radical (HO^*) and finally water H_2O. Nitric oxide (NO^*) may also form in the metabolic processes. All respiring cells produce free radicals in chloroplasts, mitochondria, endoplasmic reticulum, peroxisomes and glyxysomes. ROS may harmful for biological systems because they cause oxidative stress, the generations and accumulation of ROS which damage cell membranes, lipids, proteins, and DNA. At the same time ROS can also act in signal transduction and participate in the formation of intermediates (prostaglandins, leikotriens, etc.), they play an important role for the expression of several transcription factors (heat schok-inducing factor, nuclear factor, the cell-gene p53, nitrogen-activated protein kinase, and etc). Oxidative stress also plays a role in apoptosis in two pathways, the death-receptor and the mitochondrial (Winston, 1990, 1991; Winston & Di Giulio, 1991; Livingstone, 2001; Burlakova, 2005; Lesser, 2006; Vladimirov & Proscurina, 2007).

Antioxidant system plays the key role in inactivation of reactive oxygen species (ROS) and thereby control oxidative stress as well as redox signaling. Antioxidant defense of organism depends on many abiotic (temperature, season, salinity, oxygen concentration) and biotic (feeding conditions, food quality and quantity) environmental factors (Martinez-Alvarez et al., 2005). Organism physiological status is also influences on antioxidant defense (Rudenva et al., 2010a). Whereas these aspects have been amply studied in mammals and especially in humans, information on fish and other aquatic animals is very limited.

Antioxidant defense of the organism is detected by interactions between prooxidant and antioxidant processes and the level of the antioxidants (Figure 1) (Burlakova, 2005). Antioxidant system presents in all living organisms and it includes low molecular weight scavengers and special adopted enzymes.

Low molecular weight antioxidants are non-enzymatic compounds such as vitamins A,E,K and C, carotenoids, SH-containing substances (glutathione) and small-molecule antioxidants such as uric acid, urea, etc (Winston, 1991). Ascorbic acid functions as reductant source for many ROS, it scavenges both H_2O_2 and O_2^-, HO*, lipid hydroperoxides without enzyme catalysts. Additionally it plays an important role in recycling α –tocopherol to its reduced form (Lesser, 2005; Li et al., 2006). Glutathione(GSH) is a tripeptide (GLU-CYS-GLY) which forms a thiol radical that interacts with a second oxidized glutathione and forms a disulphide bonds (GSSG). The ratio GSH/GSSG is detectable as biomarker of oxidative stress in the organism (Livingstone, 2001; Papp et al., 2007).

GSH oxidizes H_2O_2 and organic hydroperoxides in oxidized form GSSG spontaneously or enzymatic by glutathione peroxidase, then GSSG reduces again to GSH according the equation

$$\text{GR-GSSG} + \text{NADPH}^+ + \text{H}^+ \rightarrow 2\text{GSH} + \text{NADP}^+$$

Tocopherols, especially α –tocopherol (vitamin E) are lipid-soluble antioxidants which are located within the bilayers of cell membranes and protect them against ROS (Burlakova, 2005). It is multifunctional antioxidant which is the result of its ability to quench both $^1O_2$ and peroxides. Carotenoids are also lipid-soluble compounds that protect against $^1O_2$ because they have highly conjugated double bonds and they quench ROS and can prevent lipid peroxidation in marine animals (Winston, 1990; Hussein et al., 2006). Additionally carotenoids are the precursors of vitamin A which can be a potent antioxidant (Lali & Lewis-McCrea, 2007). Vitamin K also can protect the organism against ROS (Hardy, 2001).

Uric acid can quench both $^1O_2$ and $HO^{\cdot}$. It is found in high concentrations in marine invertebrates and in can be a potent antioxidant (Lesser, 2006). There are some others low molecular weight antioxidants in aquatic organisms but in some cases their nature is unknown.

Enzymatic antioxidant system includes some enzymes which catalyze the reactions of ROS degradation. Superoxide dismutase (SOD) (EC 1.15.1.1.) protects against oxidative damage by catalyzing the reaction of dismutation of the superoxide anion to H_2O_2 :

$$2O_2^{\cdot} + 2H^+ \rightarrow H_2O_2 + O_2$$

Catalase (CAT) (EC 1.11.1.6.) catalyzes the reduction of hydrogen peroxide to water

$$2H_2O_2 \rightarrow 2H_2O + O_2$$

Selenium-dependent glutathione peroxidase (SeGPX) (EC 1.11.1.9) catalyzes the reaction of hydroperoxides reduction:

$$H_2O_2 + 2\ GSH \rightarrow 2\ H_2O + GSSG$$

Total glutathione peroxidase (PER) (the sum of SeGPX and selenium-independent GPX activities) reduces both hydrogen peroxide and organic hydroperoxides also:

$$ROOH + 2GSH \rightarrow ROH + H_2O + GSSG$$

Glutathione reductase (GR) (EC 1.6.4.2.) catalyzes the reaction

$$NADPH + H^+ + GSSG \rightarrow NADP^+ + 2GSH$$

and maintains a ratio GSH/GSSG under oxidative stress.

Quinon oxidoreductase (NAD(P)H-dependent DT-diaphorase (EC 1.6.99.2) catalyzes the two electron reduction of redox cycling quinones and related compounds to hydroquinones so preventing their univalent reduction to quinine anion radicals leading ROS production via autooxidation (Peters & Livingstone, 1996).

$$NAD(P)H + Q + H^+ \rightarrow QH_2 + NAD(P)^+$$

where Q and QH_2 are respectively quinoneand hydroquinone.

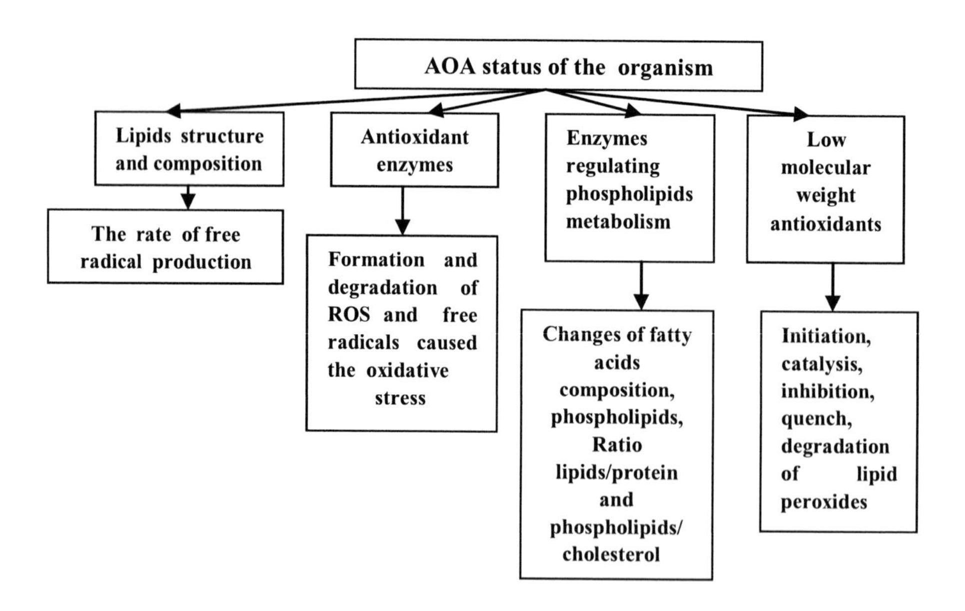

Figure 1. The links between prooxidant and antioxidant processes in the organism (adopted from Burlakova, 2005).

Previously we described the variations in blood antioxidant system of some Black Sea elasmobranch and teleosts which reflected adaptive strategy of fish species to oxidative stress and their ability to cope with the environment (Rudneva, 1997a). Other researchers reported also that antioxidant status can be correlated with animal phylogenic position and ecological peculiarities and these parameters in primitive life are less than in mammals and birds (Rocha-e-Silva et al., 2004; Sole et al., 2009). High contents of low molecular weight antioxidants including vitamins E, K, C in organs and glutathione in red blood cells in marine elasmobranch have been found and these substances might compensate for the low level of their enzymatic activity (Dafre & Reischl, 1990; Rudneva, 1997a).

At present the anthropogenic press on natural ecosystems especially aquatic locations led high level pollution. Xenobiotics distributed in water and accumulated in sediments and biota stimulate oxidative stress in organisms and modify the interactions between the main metabolic processes which influences negatively on organism. Thereby the investigations of antioxidant system is very important for understanding defense mechanisms against pollution in fish of various taxonomic and ecological groups. Additionally it's important for aquaculture purposes for development the optimal conditions of fish production and for receiving the information of repair mechanisms and adaptation to unfavorable impact. In our previous studies we showed the modification effects of anthropogenic pollution on antioxidant status of some Black Sea fish teleosts (Rudneva, 1998a,c; Rudneva et al., 2005; 2008).

Thus, comparative study of fish ROS detoxification metabolism and resistance to oxidative stress led the anthropogenic pollution is very important for the understanding the different strategy of adaptations of aquatic organisms and for risk assessment. In this case the analysis of fish antioxidant system of different taxonomic and ecological groups allows to obtain the additional information of adaptive mechanisms development in aquatic vertebrates and invertebrates and their resistance to environment changes.

Taking it into account the main objectives of the present work were the following:

- to study the antioxidant status in red blood cells, muscle, liver and gonads in different Black Sea fish species;
- to compare antioxidant defense in teleosts and elasmobranches;
- to evaluate the dependence of enzymatic activity from fish ecological characteristics

2. MATERIALS AND METHODS

2.1. Animals

The following fish species were used: elasmobranch dog fish *Squalus acanthias* .(n=12) and teleosts: horse mackerel *Trachurus mediterraneus* (n=6), high body pickarel *Spicara smaris* (n=8), round goby *Neogobius melanostomus* (n=8), scorpion fish *Scorpaena porcus* (n= 12), red mullet *Mullus barbatus ponticus* (n =10), whiting *Merlangus merlangus euxinus* (n=6), flounder *Platichthys flesus luscus* (n=5), Black Sea turbot *Psetta maxima maeotica* (n=12).

Animals were caught in spring-summer period 1998-2002 from Sevastopol Bay in Sevastopol region (Black Sea, Ukraine) (Figure 2). Fish were transported to the laboratory in the containers with marine water and constant aeration. Tested fish were classed in four groups: pelagic (*T. mediterraneus ponticus)*, suprabenthic/pelagic (*S. acanthias*, *S. smaris* and *M. merlangus euxinus)*, suprabenthic (*M. barbatus ponticus)* and benthic (*P. flesus luscus, P. maxima maeotica, N. melanostomus* and *S. porcus)* (Table 1).

2.2. Sample Preparation

Blood was taken by caudal arteria puncture and serum was separated. Red blood cells were processed as we described previously (Rudneva, 1997a). The sediment was washed three times with 0.85% NaCl solution and then lysed by addition of 5 vol of distilled water for 24 h at the refrigerator. The enzyme activity was determined in the lysates immediately after preparation.

Fish were dissected and the liver, muscles and gonads (maturation stage IV-V, separately from male and female) were quickly removed at the ice. The organs were washed in the cold 0.85 % NaCl solution several times, and then homogenized in the physiological solution used glass homogenizer. The resulting homogenates were centrifuged at 5000 g for 20 min. The supernatants were used for further enzyme analysis.

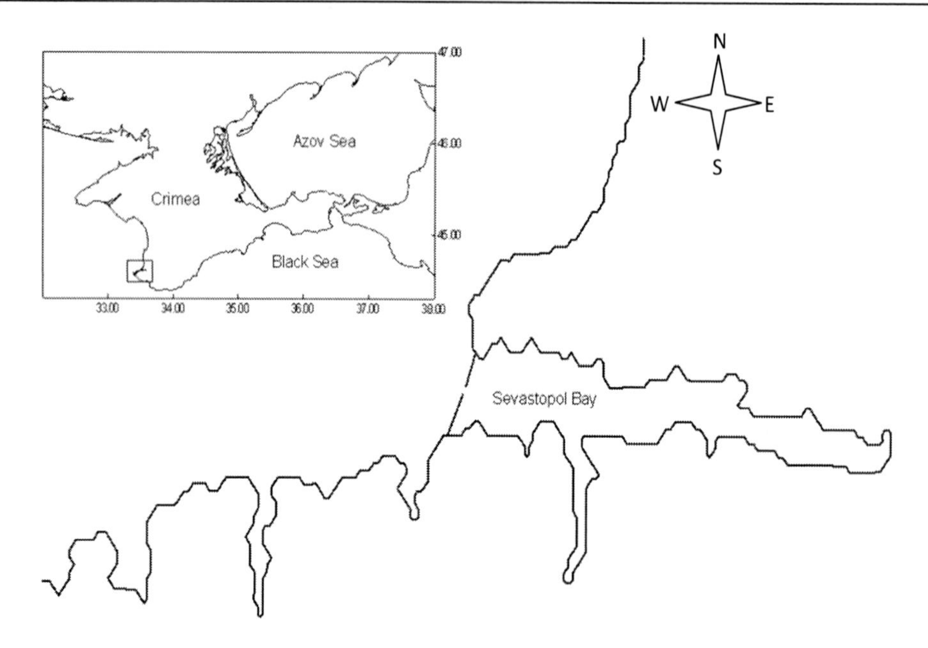

Figure 2. Sampling sites of fish specimens in Sevastopol Bay (Sevastopol, Black Sea, Ukraine).

2.3. Enzymatic Activities Assays

Antioxidant activities in fish blood cells were determined according the methods which we described previously with small modifications (Rudneva, 1997a).

Superoxide dismutase (SOD, EC 1.15.1.1) was assayed on the basis of inhibition of the reduction of nitroblue tetrasolium (NBT) with NADH mediated by phenazine methosulfate (PMS) under basic conditions (Nishikimi et al., 1972). All measurements were performed in 0.017 M sodium pyrophosphate buffer pH 8.3 at +25°C. The reaction mixtures contained 5 μM NBT, 78 μM NADH, 3.1 μM PMS, and a 0.1 ml sample; the final volume was 1.5 ml. The reaction was carried out in the cuwett of spectrophotometer at 560 nm.

Catalase (CAT, EC 1.11.1.6) was measured by the method involving the reaction of hydroperoxide reduction.

Peroxidase (PER, EC 1.11.1.7) activity was detected by spectrophotometric method using benzidine reagent (Litvin, 1981). The reaction mixture contained 1 ml acetate buffer pH 5,4, 0.4 ml 0.09% benzidine, 0.2 ml 0.03% H_2O_2, and 0.2 ml sample. The reaction followed in a spectrophotometer for 1 min at 20°C and at 600 nm.

Glutathione reductase (GR, EC 1.6.4.2) activity was assayed spectrophotometrically using a method modified after Goldberg and Sparner (1987). The reaction mixture contained 0.1 ml mM NADPH, 0.5 ml 7.5 mM oxidized glutathione, 0.2 ml mM EDTA, and 2 ml 0.05 M phosphate buffer pH 8.0. After incubation for 10 min, the extinction of the mixture was determined at 340 nm.

The enzyme activities were estimated per ml per min in the case of the fish blood and per mg protein per min in the case of the tissues. The protein concentration in the liver homogenates was estimated by the method of Lowry et al., (1951) using human serum albumin as the standard protein.

2.4. Low Molecular Weight Antioxidants Determinations

Glutathione

The content of glutathione was determined spectrophotometrically using the alloxan reagent at the 305 nm (Putilina, 1982). The content of glutathione in the sample was calculated on the graphical determination using the standard concentration of the glutathione as we described previously (Rudneva, 1997a).

Serum SH-groups

Concentration of SH-groups (total, protein and non-protein) was measured by spectrophotometric method, based on the interaction between molecular iodine and free SH-groups of the proteins and non-protein substances (Folomeev, 1981). The extinction was measured at 500 nm after 10 sec from the start. The content of non-protein SH-groups was estimated in supernatant after protein sedimentation by 3% TCA solution. The concentration of protein SH-groups was estimated as a difference between the concentration of the total and non-protein compounds (Rudneva, 1997a).

Table 1. Ecological characteristics of examined Black Sea fish species (Svetovidov, 1964)

Fish ecological group			
Benthic	Suprabenthic	Suprabenthic/pelagic	Pelagic
Scorpion fish *Scorpaena porcus* L (slow swimming, predator)	Red mullet *Mullus barbatus ponticus* Essipov (sluggish, carnivorous)	Dog fish *Squalus acanthias* L. (fast swimming, predator)	Horse mackerel *Trachurus mediterraneus* Staidachner (fast swimming, plankton feeder)
Round goby *Neogobius melanostomus* Pallas (slow swimming, carnivorous)		Whiting *Merlangus merlangus euxinus* Nordmann (less mobile, predator)	
Solea			
Flounder *Platichthys flesus luscus* (slow swimming, carnivorous)		High body pickarel *Spicara smaris* Rafinesque (less mobile, omnivorous)	
Turbo *Psetta maxima maeotica* (slow swimming, carnivorous)			

Vitamin A and Carotenoids

The contents of carotenoids and vitamin A were assayed spectrophotometrically in the lipid extracts of the tissues obtained by the homogenization in hexan:isopropanol mixture 2:1

(v/v) in cold conditions. The content of vitamin A was measured at 325 nm and the carotenoids at 450 nm respectively (Karnauhov & Fedorov, 1982).

Vitamin E

The content of vitamins was assayed by the method described Filipovich et al. (1975). For vitamin E determination the tissue extracts were hydrolyzed in a mixture of 60% KOH and 96% ethanol in ratio 5:4 (v:v). Then tocopherol was extracted by diethyl ether which was evaporated and the sample was solved in the ethanol. After oxidation by concentrated HNO_3 and boiling during 3 min the absorbance was measured at 470 nm.

Vitamin K

Vitamin K determination was detected spectrophotometrically according the method described Filipovich et al. (1975) with small modifications (Rudneva, 1999). The absorbance was assayed at 448 nm and the content was calculated from a standard curve of known concentrations of vitamin K.

Antioxidant Activity Assay

Antioxidant activity was measured in lipid extracts prepared as described above by fluorescent method in Luminoscan (LKB, Sweden). The results were estimated in arbitrary units.

2.5. Statistical Analysis

The results were processed to statistical evaluation with Student's tests for each paired sample. All numerical data are given as means ± SD (Halafyan, 2006). The significance level was <0.05. The correlation coefficients were calculated by the least-squares method between enzyme activities and between the average values of enzymatic activities of every examined fish species.

3. RESULTS

3.1. Blood Antioxidant System

The obtained results showed the differences between the examined parameters in the blood of the Black Sea fish species. The concentrations of vitamin A in serum of the tested fish species varied insignificantly from 4. 7 to 9.1 mg g^{-1} lipids (Figure 3). The highest concentration was shown in *N. melanostomus*, in other fish species vitamin A content was the similar. Glutathione concentration in red blood cells of the fish ranged between 22.8 and 60.0 μg% (Figure 4). The highest values were indicated in *N. melanostomus*, *T. mediterraneus* and in *P. flesus lucsus*, the least values were shown in *S. acanthias* and *M. merlangus euxinus*.

The concentration of serum SH-groups is presented in Figure 5. Content of the total SH-groups was significantly higher in the serum of *S. acanthias* ($p<0.01$) as compared with the teleost fish species. Among the teleosts high values were indicated in *S. porcus*, *S. smaris* and

M. merlangus euxinus while in other species they were less. In all tested teleost fish concentration of protein SH-groups was higher than the content of non-protein, in elasmobranch the opposite trend was marked: in *S. acanthias* content of non-protein SH-groups was greater than the protein ones. The ratio of protein and nonprotein SH-groups was also differed among examined fish species (Table 2).

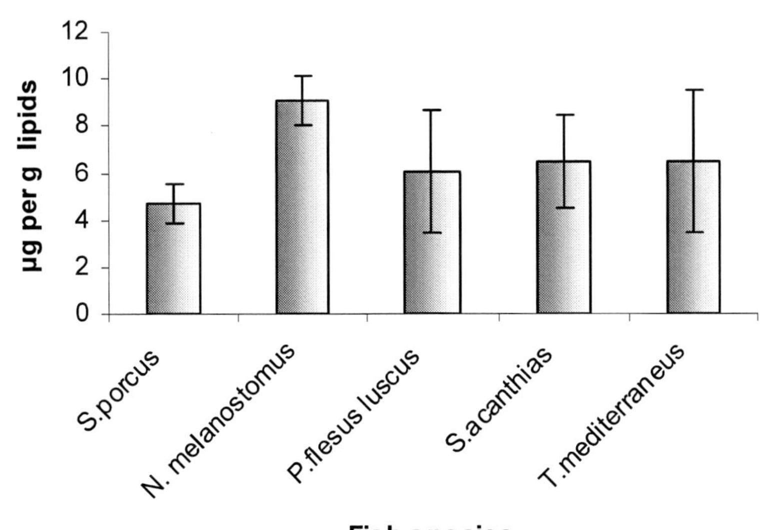

Figure 3. Vitamin A content in the serum of some Black Sea fish species (mean ± SD).

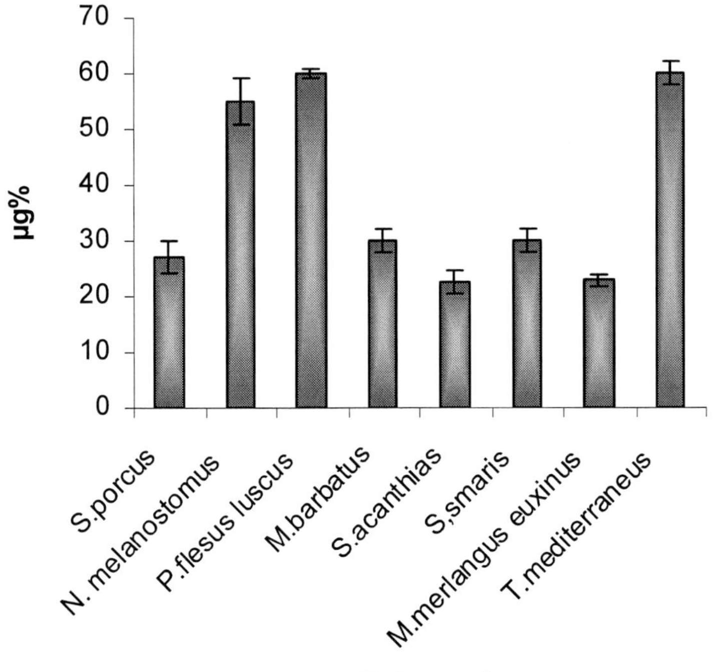

Figure 4. Glutathione content in red blood cells of Black Sea fish species (mean ± SD).

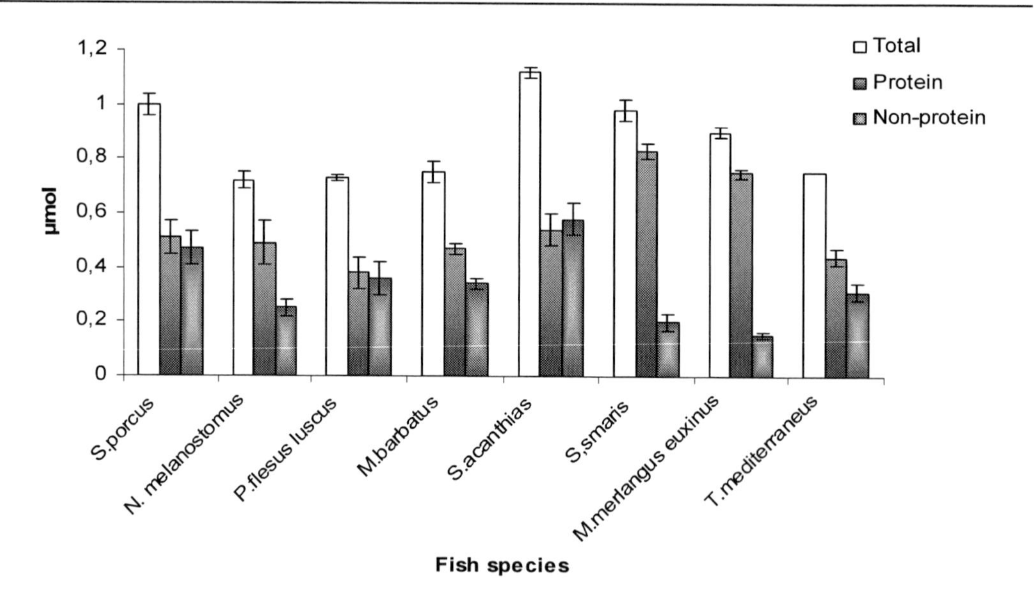

Figure 5. Concentrations of SH-groups in blood serum of Black Sea fish species (mean± SD).

Table 2. Ratio of protein and glutathione concentrations, protein and non-protein SH-groups content in blood serum of Black Sea fish species

Fish species	Protein concentration, mg%, mean ±SD	Ratio protein/ non-protein SH-groups	Ratio protein SH-groups/ protein concentration	Ratio non-protein SH groups/ glutathione concentration
S. porcus	5.21 ±1.2	1.13	0.10	0.020
N. melanostomus	2.22 ±0.6	2.03	0.22	0.004
P. flesus luscus	1.20 ±0.4	1.08	0.31	0.020
M. barbatus	3.95 ±1.0	1.20	0.10	0.010
S. acanthias	2.99±0.6	0.93	0.18	0.025
S. smaris	1.24 ±0.4	3.04	0.59	0.008
M. merlangus euxinus	1.30 ±0.3	5.00	0.58	0.006
T. mediterraneus	3.20 ±0.7	1.42	0.14	0.005

Protein concentration in fish blood serum ranged between 1.20 and 5.20 mg% and demonstrated interspecies differences. Factor protein/non-protein SH-groups was higher in the fish belonging to suprabenthic/pelagic group with the exception of elasmobranch *S. acanthias* which was the least and <1 as compared with the teleost species. The ratio protein SH-groups/protein concentration showed the great values in suprabenthic/pelagic fish than in others including elasmobranch belonging to the same ecological group. Ratio non-protein SH groups/glutathione concentration was the highest in *S. acanthias*, in teleosts it varied widely from 0.004 to 0.020.

Antioxidant enzyme activities differed in red blood cells of examined fish species (Figure 6). SOD activity was low in *S. acanthias* and *M. barbatus ponticus* ($p<0.01$) than in other teleost species. High enzymatic activity indicated in *M. merlangus euxinus, S. smaris* and *T. mediterraneus.* CAT activity was not shown in elasmobranch blood, while in teleosts it was low with the exception of *M. merlangus euxinus* and *T. mediterraneus.* The greatest PER activity was identified in *M. merlangus euxinus,* the least in *P. flesus luscus* and *M. barbatus ponticus.* GR activity varied less in the blood of tested fish species both elasmobranch and teleosts. The high level was in *M. barbatus ponticus* and in *M. merlangus euxinus,* the least in *S. smaris.* Thus the obtained results demonstrated the interspecies differences in blood antioxidant system of examined fish depending on their phylogeny and ecology.

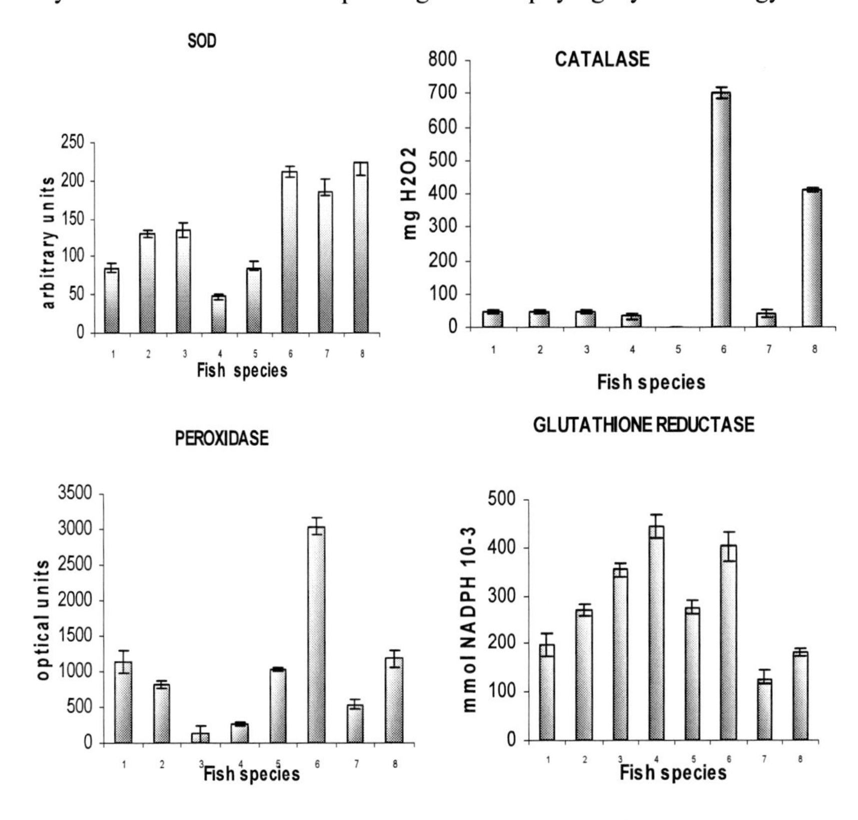

Figure 6. Antioxidant enzyme activities in Black Sea fish red cells (ml $^{-1}$red cells min^{-1}, mean$\pm$ SD). 1- *S.porcus,* 2 - *N.melanostomus,* 3 - *P. flesus luscus,* 4- *M. barbatus ponticus,* 5- *S. acanthias* 6- *S. smaris,* 7 - *M. merlangus euxinus* , 8 - *T. mediterraneus.*

3.2. Muscle Antioxidant System

Glutathione concentration in fish muscle varied from 58,4 in *N.melanostomus* to 109.8 mg % in *S. acanthias* (Figure 7) . High concentration of glutathione was indicated in *M. barbatus ponticus* and it was significantly greater than in other tested teleost species ($p<0.01$).

The greatest content of carotenoids was shown in *M. barbatus ponticus* tissues (13.46 µg · g^{-1} lipids) while in other forms it varied from 2.09 µg · g^{-1} lipids in *S porcus* to 6.93 µg · g^{-1} lipids in *S. smaris* (Figure 7). Vitamin A varied insignificantly in fish tissues while vitamin E ranged between 0.37 µg · g^{-1} wet weight in *S. porcus* and 7.01 µg · g^{-1} wet weight in *M. barbatus ponticus*. The highest level of vitamin K was indicated in *S. acanthias* muscle which was significant greater than in examined teleost species. The least content of vitamin K was shown in *S. porcus* tissues.

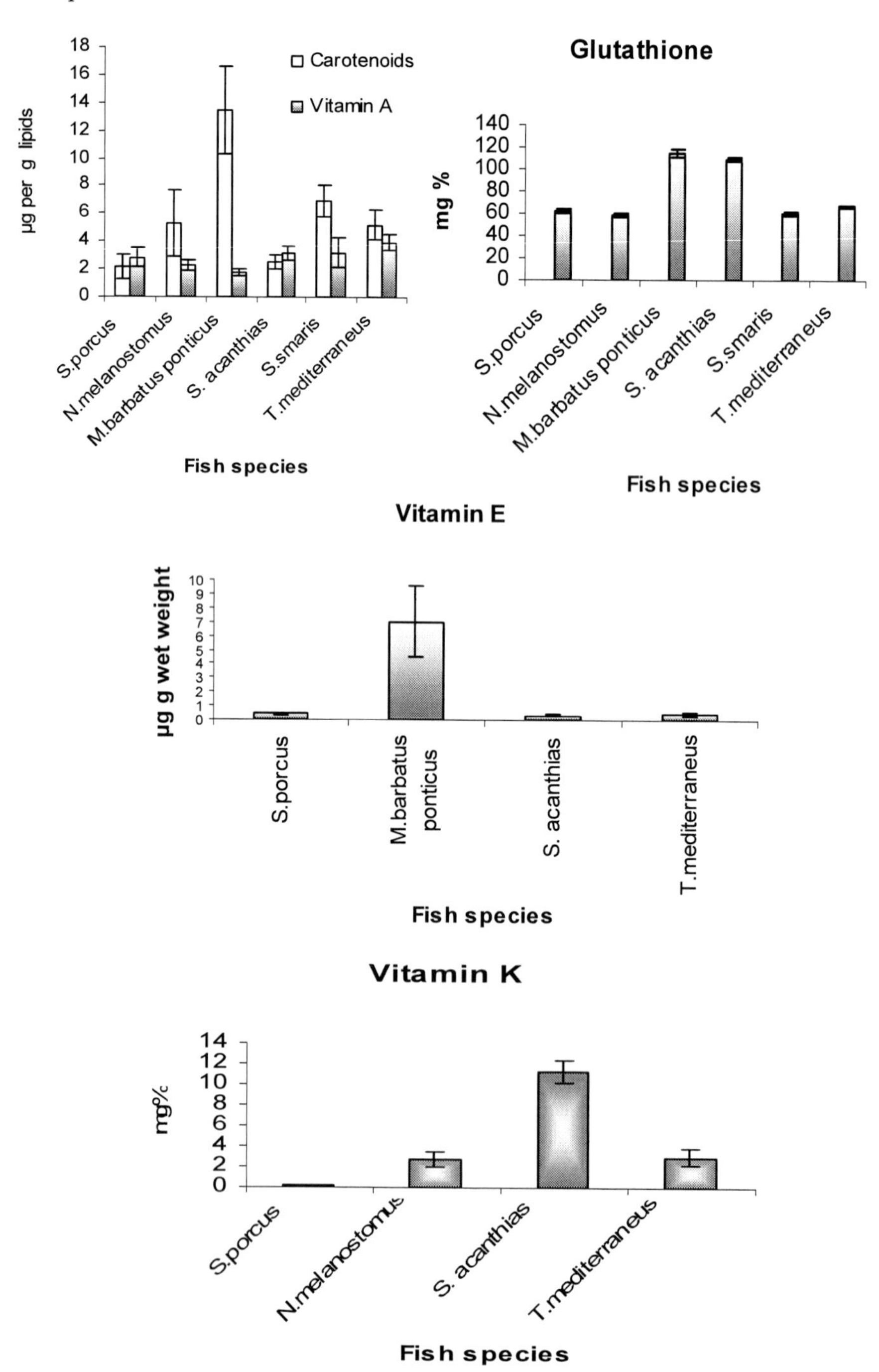

Figure 7. Content of low molecular weight antioxidants in muscle of Black Sea fish species (mean ± SD).

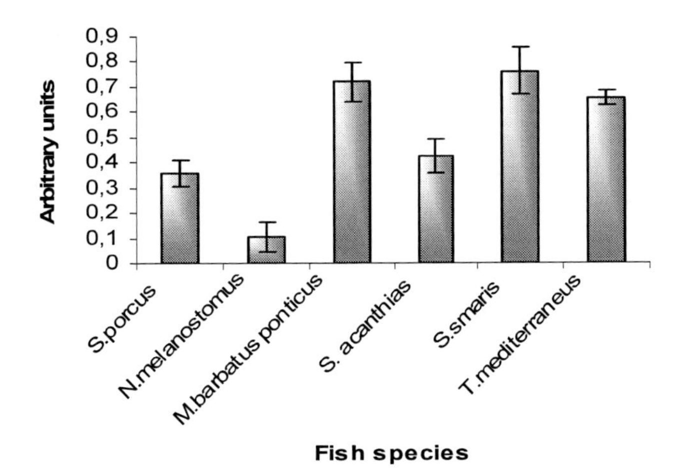

Figure 8. Total AOA of lipids in Black Sea fish muscle (mean± SD).

The total AOA of lipids in muscle was the least in *N. melanostomus* (0.105 arbitrary units), in other fish species it ranged between 0.358 in *S. porcus* and 0.762 in *S. smaris* (Figure 8).

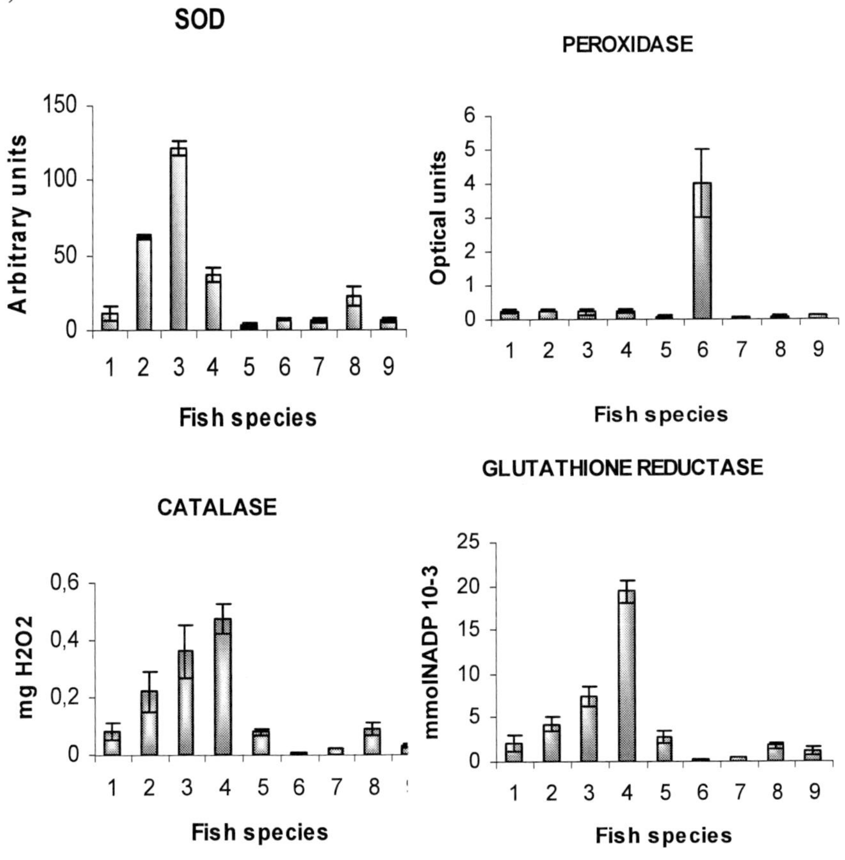

Figure 9. Antioxidant enzyme activities in Black Sea fish muscles (mg $^{-1}$protein min^{-1}, mean± SD). 1- *S.porcus*, 2 - *N. melanostomus*, 3 - *P. flesus luscus*, 4- *P.maxima maeotica*, 5- *M. barbatus ponticus*, 6- *S. acanthias* 7- *M. merlangus euxinus*, 8- *S. smaris*, 9 - *T. mediterraneus*.

SOD activity was varied in fish muscle from 39.2 to 129.8 arbitrary units .mg-1 protein .min-1 (Figure 9). The highest enzymatic activity was noted in P. flesus luscus, in other benthic forms it was also in 1.9-31.1-fold higher than in species belonging to pelagic, suprabenthic and suprabenthic-pelagic classes, in which the activity was less with the exception of S. smaris. In elasmobranch tissues SOD activity was the similar as in teleosts excepted flounders, S smaris and N. melanostomus ($p<0.02$). CAT and GR activities were low in fish muscle but in benthic species they were higher than in other forms. CAT and GR activity in elasmobranch muscle was significantly less ($p<0.01$) than in teleosts. Great enzymatic activity was found in benthic forms. PER activity was the similar in all tested teleost fish species, in S. acanthias it was the highest and more than 20-40-times greater than in teleosts ($p<0.01$). Enzymatic activity in benthic fish was less than in elasmobranch ($p<0.01$) but it was higher than in other examined forms.

Significant correlations was indicated between CAT and GR activity ($r=0.89$) and insignificant between SOD and CAT ($r=0.44$). In other cases no correlations were shown.

The obtained results demonstrated the differences between examined fish species depending on their ecological status and phylogenetic position: PER activity was unusually high in elasmobranch and SOD, CAT and GR levels were greater in benthic species as compared with fish belonging to other ecological groups.

3.3. Hepatic Antioxidant System

Glutathione content varied insignificantly in hepatic tissues of the examined fish species (Figure 10). The highest concentration of carotenoids and vitamin E was shown in *M.barbatus ponticus* liver which was much more than 4-7-times greater than in other forms. Vitamin A content varied insignificantly in fish liver tended to increase in *N. melanostomus* and *S. smaris,* the content of vitamin K was 4,5-fold greater in *S. acanthias* as compared with *S porcus*.

The least AOA of lipids was indicated in the liver of *N. melanostomus*, the highest was in *M. barbatus ponticus* and *S. acanthias* (Figure 11). The values of other fish species were intermediate.

Antioxidant enzyme activities are presented in Figure 12. SOD activity ranged between 2.95 and 115.6 arbitrary units mg^{-1} protein min^{-1}, the highest value was shown in *P. maxima maeotica* and the least in *T. mediterraneus*. CAT activity varied insignificantly in fish liver with the exception of *P. maxima maeotica* in which it was higher more than 10-fold as compared with other examined fish species. It was correlated with high level of hepatic SOD activity in this form. Great values of PER activity was indicated in *S. acanthias* and *S smaris*, in other forms it was significantly less ($p<0.01$). As in muscle the least GR activity was found in elasmobranch liver ($p<0.01$) as compared with teleosts. Among teleost species *S. porcus* demonstrated the increase of GR activity in the liver while in other species it varied insignificantly. Correlations between hepatic antioxidant enzyme activities were not shown, only insignificant link was indicated between SOD and CAT activity ($r=0.39$).

Thus in liver the interspecies differences between enzymatic and non-enzymatic antioxidants were also shown but they were not uniform as compared with the muscle ones.

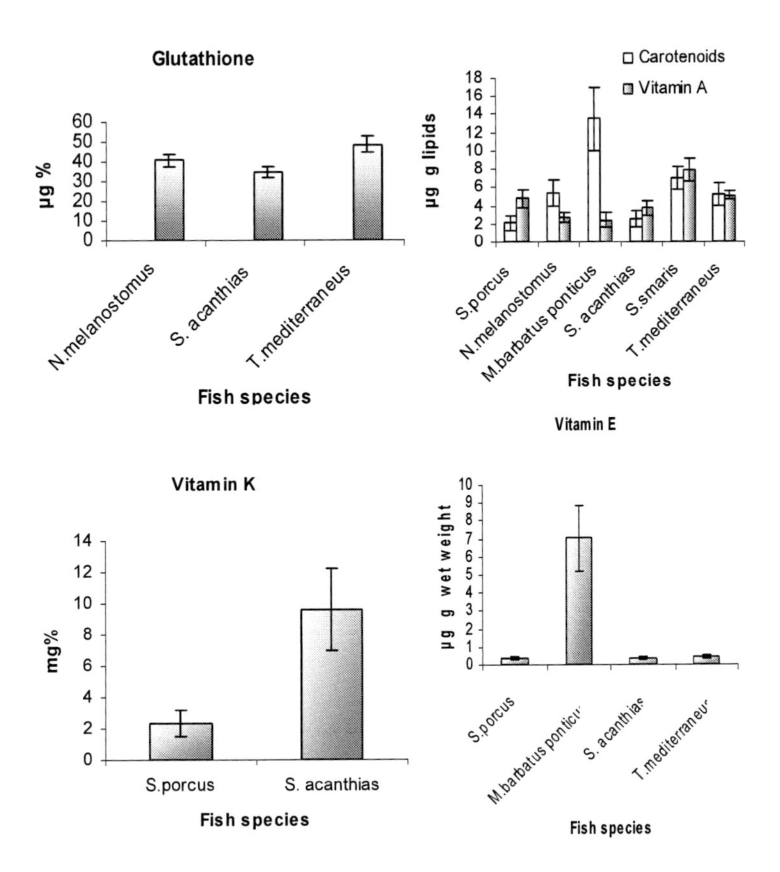

Figure 10. Low molecular weigh antioxidants content in the liver of some Black Sea fish species (mean ±SD).

3.4. Gonads Antioxidant System

Glutathione content was high in the male gonads of *N. melanostomus* and *S. acanthias* while in other tested fish it was significantly less ($p<0.01$, Figure 13).

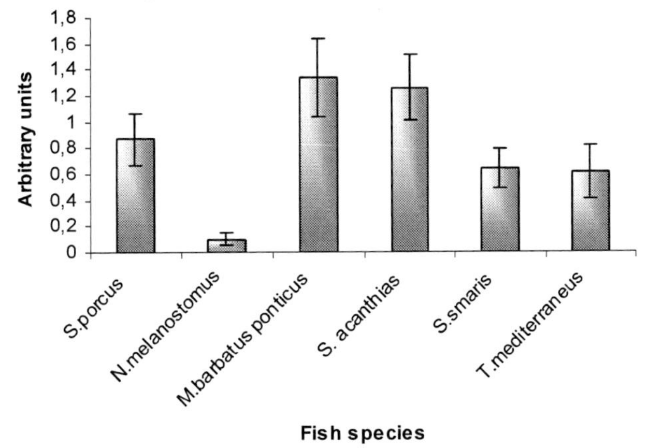

Figure 11. Total lipid AOA in Black Sea fish liver (mean ±SD).

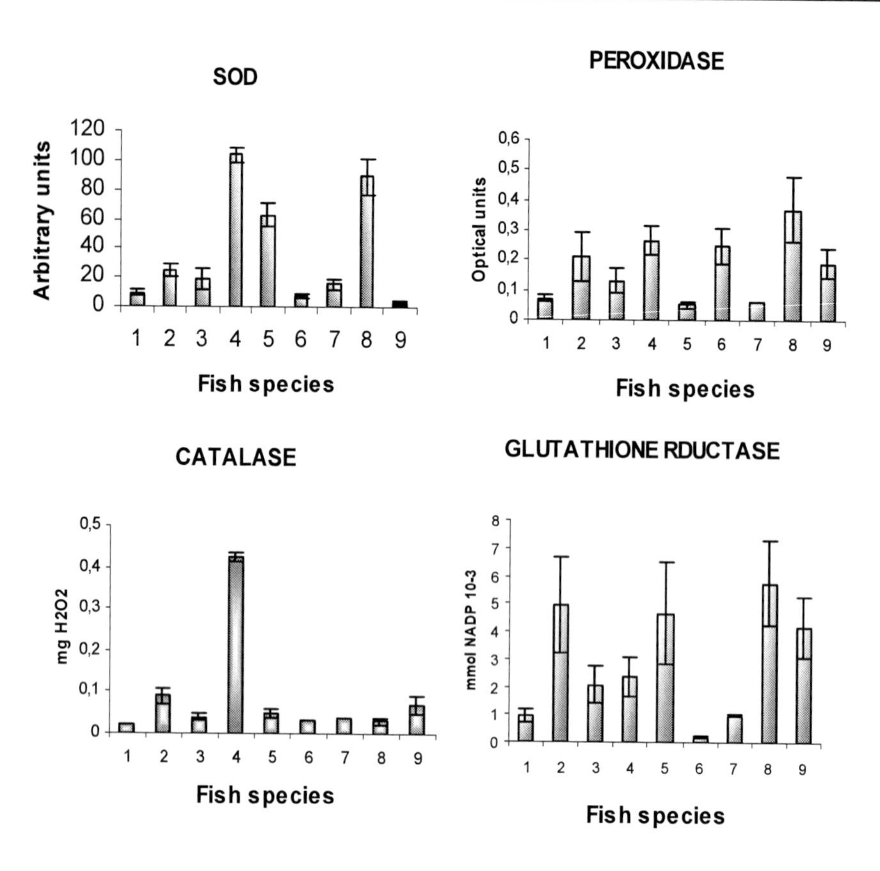

Figure 12. Antioxidant enzyme activities in Black Sea fish liver (mg $^{-1}$protein min^{-1}, mean± SD). 1- *S. porcus*, 2 - *N. melanostomus*, 3 - *P. flesus luscus*, 4- *P. maxima maeotica*, 5- *M. barbatus ponticus*, 6- *S. acanthias* 7- *M. merlangus euxinus*, 8- *S. smaris*, 9 - *T. mediterraneus*.

The greatest concentration of carotenoids was shown in male gonads of *N. melanostomus*, in other teleosts it was less. No differences were noted between male and female of *S. porcus* and *T. mediterraneus* while in male *S. acanthias* and *S. smaris* carotenoids concentration was significantly higher ($p<0.01$) than in female. Vitamin A, E and K contents varied insignificantly in the gonads of examined teleost species, the least value was detected in elasmobranch. Differences between male and female were not found.

Total AOA of lipid in fish gonads is presented in Figure 14. In *S. acanthias* it was the similar in both sex, in teleosts it was higher in females than in males. The least value was detected in the male of *N. melanostomus* and the greatest was shown in elasmobranch.

Antioxidant enzyme activities were also varied in the gonads of tested fish species (Figure 14). SOD activity ranged between 4.34 and 55.68 arbitrary units mg^{-1} protein min^{-1}. The highest value was indicated in the female *M. barbatus ponticus* and *S. smaris*, while the least one was in *T. mediterraneus* and *S. acanthias*. In elasmobranch SOD activity was less than in the majority of teleosts excepted the female of *S. porcus* and *T .mediterraneus* where they were the similar. No differences were shown between the enzymatic activities in male and female gonads in *S. acanthias* while in testes of *N. melanostomus* it was higher approximately in 7-fold as compared with ovaria.

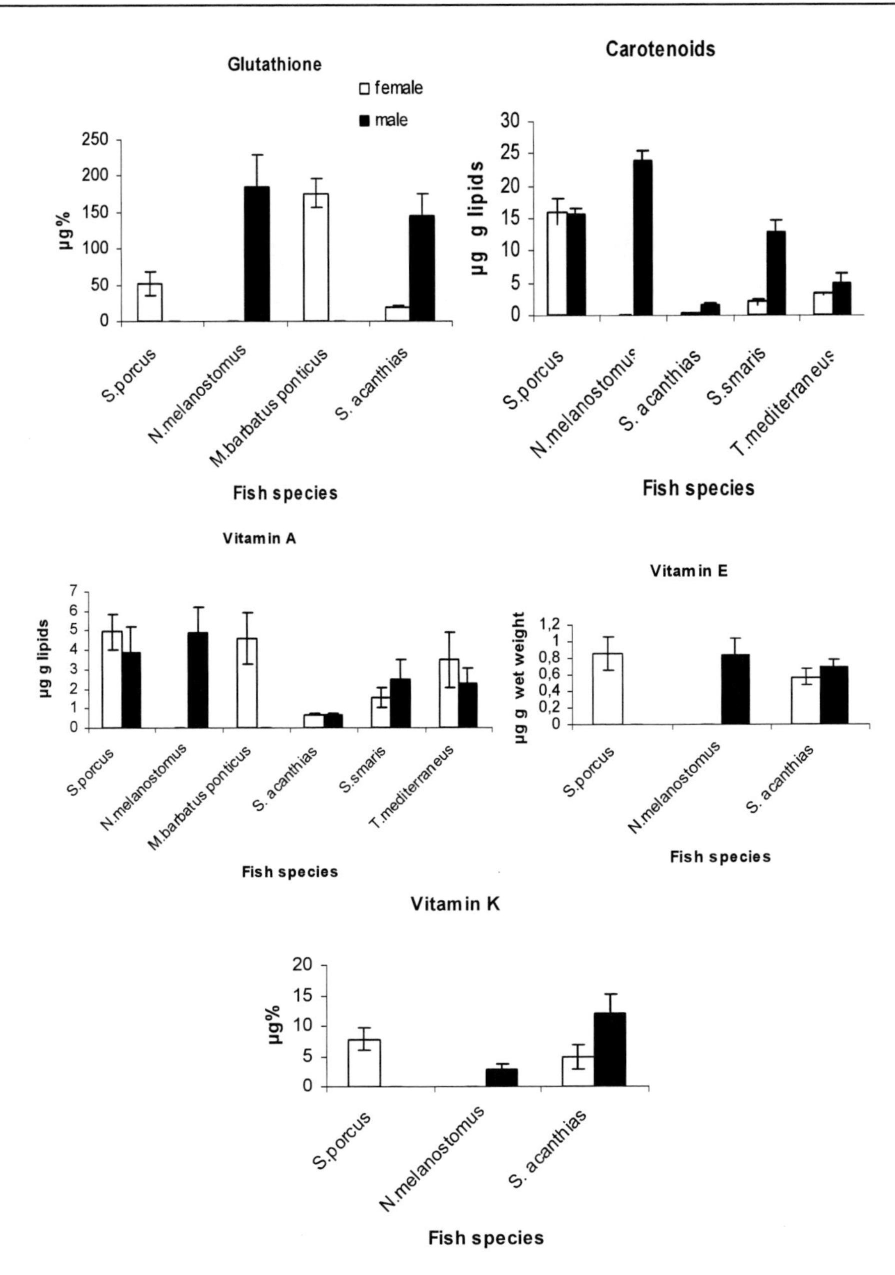

Figure 13. Low molecular weight antioxidants content in the gonads of Black Sea fish species (mean $\pm$ SD).

CAT activity varied from 0.01 to 0.29 mg H_2O_2 mg^{-1} protein min^{-1} and in female it was significantly greater ($p<0.01$) than in male in *S. acanthias* and *S. porcus*. PER level in fish gonads ranged between 0.03 and 1.78 optical units mg^{-1} protein min^{-1}. As compared with CAT, PER activity demonstrated the opposite trend in fish gonads: in male enzymatic activity was significantly higher ($p<0.01$) than in female both in elasmobranch and teleost *S. porcus*. Similar trend was shown in GR activity which was significantly greater ($p<0.01$) in gonads of females of *S. porcus* and *S. acanthias* as compared with males.

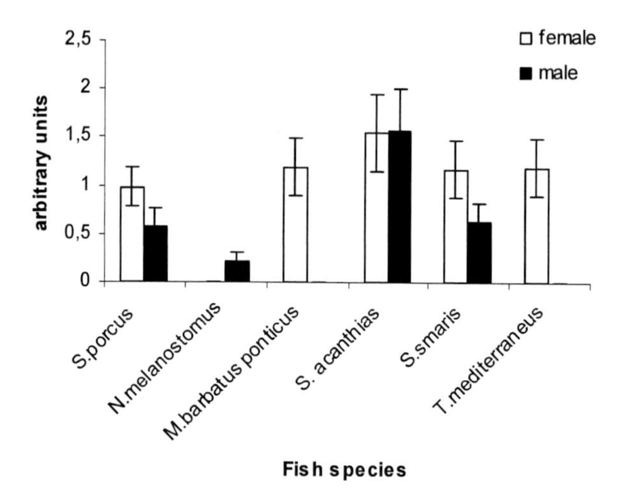

Figure 14. Total lipid AOA in the gonads of Black Sea fish species (mean ± SD).

GR activity varied from 0.04 to 7.39 NADPH nml mg^{-1} protein min^{-1}, the least value was noted in elasmobranch male gonads as compared with teleosts.

No correlations were shown between antioxidant enzyme activities in fish gonads with the exception of SOD and CAT and CAT and GR, but the links were insignificant (r=0.30 and r=0.37 correspondingly).

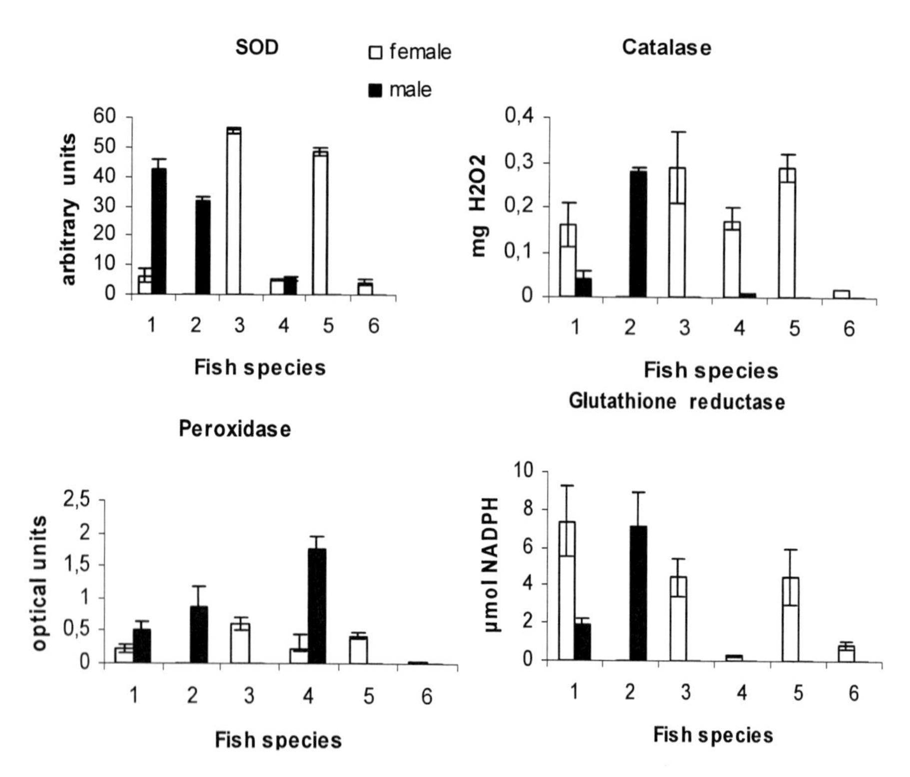

Figure 15. Antioxidant enzyme activities in Black Sea fish gonads (mg^{-1}protein min^{-1}, mean± SD). 1- *S. porcus*, 2 - *N. melanostomus*, 3- *M. barbatus ponticus*, 4- *S. acanthias* 5- *S. smaris*, 6 - *T. mediterraneus*.

Thus antioxidant enzyme activities were very low in fish gonads. At the same time they contain high concentrations of low molecular weight antioxidants which play an important role in antioxidant defense of reproductive system against ROS and oxidative stress.

4. Discussion

4.1. Tissue Localization Specificity

Antioxidant enzymes and low molecular weight scavengers were localized in all examined tissues of fish tested with the exception of CAT activity in elasmobranch blood. It agrees with the results of other investigators (Filho, 1996; Filho &Boveris, 1993; Filho et al., 1993; Sole et al., 2009) and with our previous findings (Rudneva, 1997a). At the same time the content of antioxidants and antioxidant enzymes were ranged in fish tissues.

Fish erythrocytes have been proven to provide an excellent model for study of ROS and antioxidant status because these cells possess mitochondria and nucleus and show changes that are associated to fish physiological and ecological status (Filho, 2007). Additionally, serum low molecular weight scavengers play an important role in antioxidant defense of the organism. Our findings showed that the concentration of vitamin A in blood serum and glutathione in red cells was differed approximately in 2-fold in tested fish species. The content of the total SH-groups varied less and it was significantly higher in the serum of *S. acanthias* ($p<0.01$) as compared with the teleosts and the level of non-protein SH- groups were greater than the protein ones which we documented earlier (Rudneva, 1997a).

In our study we displayed the absence of CAT activity in blood red cells of elasmobranch *S.acanthias*. Other authors also reported the lack or relatively low contents of CAT and GR activity in blood of some fish species which was led to the presence of high levels of H_2O_2 in the blood (Filho et al., 1993; Filho & Boveris, 1993). In this case the fish could get excrete of H_2O_2 through the gills directly together with NH_3. Opposite, high levels of PER activity were indicated in fish blood which could be caused to methemoglobin that is usually at high concentrations in fish erythrocytes. This could be increase the rate of NADPH oxidation and falsify glutathione peroxidase determination (Filho et al., 1993). Besides that we could proposed that the lack of CAT activity in fish blood could be compensated high PER level which reduces both H_2O_2 and organic hydroperoxides.

The correlation was shown between SOD and CAT activity in red blood cells ($r=0.50$) and it was higher in teleost species with the exception of elasmobranch ($r = 0.64$) (Table 3). Strong correlation was estimated between CAT and PER activities ($r=0.68$), in other cases the links were less.

High contents of low molecular weight antioxidants were detectable in fish muscle. The greatest levels of glutathione, carotenoids, vitamin E were indicated in *M. barbatus ponticus* while in other examined fish species the values were the similar or less. In *S. acanthias* muscle glutathione and vitamin K concentrations were detectable in high levels also. Total AOA in lipids was less than 1, and the least was shown in *N. melanostomus* which could be connected with low level of antioxidant defense protection in muscle of this animal.

Table 3. Correlations between antioxidant enzyme activities in tissues of examined fish species

Enzyme	Blood	Muscle	Liver	Gonads
SOD→CAT	0.50	0.44	0.39	0.30
SOD→PER	0.13	-0.099	-0.098	-0.05
SOD→GR	-0.003	0.10	0.14	-0.022
CAT→PER	0.68	-0.05	-0.13	-0.15
CAT→GR	-0.11	0.89	-0.13	0.37
PER→GR	-0.14	-0.01	0.05	-0.14

Antioxidant enzyme activities varied widely in muscle of examined fish species. SOD, CAT and GR activities were detectable higher level in both flounders, while PER was greater in *S. acanthias*. Significant correlations were indicated between CAT and GR activity (r=0.89) and moderate link was noted between SOD and CAT level (r=0.44) (Table 3). Correlations were not shown between other examined enzymes level in fish muscle.

Glutathione concentration in fish liver was less than in muscle, the levels of carotenoids and vitamin K were the similar or insignificantly less while the content of vitamin A was higher, vitamin E trends were not uniform. The total AOA in lipids was greater or the similar in fish liver as compared with muscle. At the case of *S. acanthias* and *M. barbatus ponticus* it was >1, in other species it was less. The obtained results demonstrated the similar trends of antioxidants content in muscle and in the liver of the totality of fish species investigated.

Hepatic antioxidant enzymes level varied widely and the comparative study the values of hepatic and muscle enzymatic activities was not shown the uniform trends. In contrast the authors documented the highest SOD and CAT levels in fish liver as in high metabolic rate organ (Filho & Boveris, 1993; Zelinski & Portner, 2000). Our findings in examined Black Sea fish species did not agree with these data. PER activity was significantly less or the similar in fish liver than in muscle, SOD, CAT and GR levels were higher in muscle in some fish species while in others they were less as related to liver. Possibly, the ratio between antioxidant enzyme activities in fish muscle and liver depends on fish taxonomic position and ecological status which we try to discuss below.

Fish gonads contain high concentrations of low molecular weight scavengers which were greater than in muscle and liver. The levels of glutathione, carotenoids and vitamin A were significantly higher than in other tested tissues and they prevent the eggs and maturation gonads against oxidative stress. High total AOA of lipids was shown also and in female it was greater than in male with the exception of *S. acanthias* where the values were the similar. This was agree with the data of other investigators who reported of relatively high concentration of vitamin E (130 pmol/mg protein) which may constitute a protection mechanism for the environment disturbance upon fertilization which will exposed the eggs directly contact with oxygen and free radicals production (Filho et al., 1993). At the same time SOD and GR activities in the gonads of majority of Black Sea fish species were less than in muscle and in the liver.

Fish gonads and maturated eggs demonstrated exceedingly low respiration level when retained at the sac prior to spawning, a period which coincides with a fat and carotenoids transfer to the gonads and in this case low molecular antioxidants play an important role in antioxidant defense.

In *S. acanthias* gonads SOD activity was the similar in both sex, CAT and GR levels were higher in female and PER in male. Among teleosts the least enzymatic activity was shown in *T. mediterraneus* gonads belonging to pelagic group. Additionally the differences between antioxidant enzyme activities were shown also but they were not uniform. In the case of *S. porcus* SOD and PER activity was less in ovaria than in testes, CAT and GR demonstrated the opposite trends.

Thus the results obtained showed the differences between both enzymatic and non-enzymatic antioxidant levels in tissues of examined fish species associated with the specificity of metabolic rate of the organ, phylogenetic and ecological status of the animal.

4.2. Antioxidant Status and Fish Phylogenetic Position

Elasmobranch group is an ancient fish which are characterized by some particular features such as the lack of biochemical apparatus for bone formation and unusually high urea content in tissues (Filho & Boveris, 1993; Blazer V.S. 1992; Lali et al., 2007). Thus the comparative studies of antioxidant defense of elasmobranhs and teleosts are very important for understanding the main ways of evolutionary process and the mechanisms of adaptation to aerobic life. It has been reported that antioxidant system can be correlated with fish taxonomic position and there are some evidences suggesting this opinion (Filho & Boveris, 1993; Filho et al., 1993; Filho, 1996; Sole et al., 2009).

At our previous and present studies we showed the lack of CAT activity in red blood cells in elasmobranch *S. acanthias* related to all examined Black Sea teleost fish species. SOD level was also less than in major of teleosts. At the same time ratio of non-protein SH groups/glutathione concentration was the highest in elasmobranch blood as compared with teleosts. It suggests our previous proposal that the low molecular weight antioxidants especially SH-containing components play an important role in the antioxidant defense in elasmobranch and compensate the absence of CAT and relatively low SOD and GR activities. High concentration of glutathione was detected in muscle of *S. acanthias* also and vitamin K in muscle and in the liver. Enzymatic activity was less in elasmobranch tissues as compared with the teleost species excepted PER activity in muscle which was the greatest among the all tested fish.

A few comparative studies of hepatic antioxidant defense enzymes in teleosts and elasmobranchs showed that their content in primitive fish species was less than that of teleosts and seems to follow the overall metabolic oxygen consumption or activity level from each fish major taxonomic group (Filho & Boveris, 1993; Filho et al., 1993). Decrease of antioxidant enzymes level and high low molecular weight scavengers content in elsmobranch tissues related to teleosts were reported by other investigators (Martinez-Alvarez et al., 2005; Sole et al., 2009). Thus our findings provide that more primitive species such as elasmobranch have more primitive antioxidant system in which non-enzymatic compounds play an important role in protection organism against oxidative stress. At the other hand elasmobranch tissuse are rich high polyunsaturated fatty acids which are the main substrates for oxidation (Filho, 2007; Vladimirov & Proskurina, 2010) and in this case the use of low molecular weight scavengers could be more advantage in defense mechanisms. High levels of non-enzymatic antioxidants were also shown in marine invertebrates and some primitive fish species (Dabrowski & Ciereszko, 2001; Torrisen & Christiansen 1995).

In addition in our previous studies we documented the presence of great contents of lipid peroxidation products in *S. acanthias* tissues as compared with teleosts (Rudneva, 1997b; 1998b) which were the evidence of the lack antioxidant system in elasmobranch tissues. It agrees with the findings of other authors which indicated high concentrations of lipid peroxides and TBA-reactive components in elasmobranch (Filho & Boveris, 1993). Our further comparative studies of glutathione-S-transferase (GST) which plays an important role in detoxification process by catalyzing the conjugation of tripeptide glutathione with some endogenous toxic metabolites and many environmental contaminants demonstrated that enzymatic activity in elasmobranch blood was significantly less than in teleosts (Rudneva et al., 2010b). The other biotransformation enzyme EROD was also displayed the decrease of its activity in elasmobranch related to teleosts (Sole et al., 2009).

We could note some peculiarities of antioxidant defense in elasmobranch gonads. In male gonads of *S. acanthias* SOD activity was less in 6-8 fold, CAT activity in 4-28-fold and GR activity in 47-186 –fold than in teleosts while the content of non-enzymatic antioxidants was insignificantly higher or the similar as in teleosts gonads. This was only partly reflected the dominating of low molecular weight scavengers in elasmobranch gonads and we could not postulate that they significantly differed as compared to teleosts. *S. acanthias* is viviparous species and its embryo develops in maternal organism which protects against oxidative stress of environment during embryogenesis. In addition as we noted previously elasmobranch eggs contain high contents of non-enzymatic antioxidants, especially fat-soluble vitamins (Rudneva, 1999).

The main trends of antioxidants levels were more clearly in female gonads as compared with male. In elasmobranch gonads high concentrstions of glutathione and vitamin K were detected. Additionally both in ovaries and in testes we found high value of the total lipid AOA which was probably associated with the presence of fat-soluble antioxidants of unknown nature. Despite of the antioxidants content in gonads of elasmobranchs and teleosts varied widely we could note that the concentrations of fat soluble compounds were higher in the gonads of teleosts than in elasmobranch and tended to growth as following: *S acnthias* → *T. mediterraneus* →*S.smaris* → *M. barbatus ponticus* → *N. melanostomus* → *S. porcus*.

The obtained results have been shown that despite of the general trend of lower enzymatic activities in elasmobranch in relation to teleosts PER activity was unusually great in S. *acanthias* muscle and male gonads. We could suggest that PER is non-specific enzyme as compared with CAT because it reduces both H_2O_2 and organic hydroperoxides (see section Introduction) that could compensate the lack of CAT. In general our findings demonstrated that antioxidant enzyme levels in fish gonads were significantly less or tended to decrease in elasmobranch as compared with teleosts.

Thus taking into account specificity of metabolic pathways in elasmobranch as compared with teleosts we could propose the following compensatory mechanisms in their antioxidant defense:

- in blood: high concentration of SH-groups especially non-protein;
- in muscle: high concentration of glutathione and vitamin K, high level of PER which compensates low level of CAT;
- in liver: high concentration of vitamin K and high value of total lipid AOA associated with the presence of fat-soluble antioxidants of unknown nature;

- in gonads: high levels of glutathione, vitamin K and E and total lipid AOA in male; development of the embryo in maternal organism which protects it against oxidative stress of the environment.

Despite of this we could postulate that ecological peculiarities of fish species play more important role in antioxidant defense than phylogenetic ones.

4.3. Antioxidant Enzyme Activities and Fish Ecological Status

The present study supports the great interspecies differences in antioxidants activity in various tissues between examined species, which agrees with the results of other researchers, who reported about the high variability of antioxidants content in fish (Filho et al., 1993; Sole et al., 2009; Winston, 1991). Our findings demonstrated that the interspecies differences in blood, muscle and liver were more significantly than in gonads which showed more homogenous response.

In fish blood glutathione content was relatively higher in pelagic active fish *T. mediterraneus* and in sluggish benthic forms while in other species belonging to suprabenthic and suprabenthic/pelagic groups the values were intermediate. Opposite trend was shown for the total SH-groups content in serum which was less in benthic and pelagic fish. We could propose that the specificity of metabolism of active and sluggish species could influence on the SH-containing compounds which concentration correlated with fish moving activity and ROS formation. High concentration of SH-groups in blood serum of *T. mediterraneus* could be associated with the defense mechanisms against ROS production which increases in active animals. The key antioxidant enzymes such as SeGPX, GR and GST contain glutathione in their active center (Papp et al., 2007). Contrary, high concentration of glutathione in blood of benthic fish could be associated with the great levels of pollutants in bottom environment where xenobiotics concentrated in the sediments and directly and indirectly (via food chain) affected benthic biota (Sole et al., 2009).

SOD activity was significantly less in sluggish forms as compared with fast moving ones. It may indicate that the oxygen transport capacity of fish blood follows the same evolutionary trend of SOD level concentration in blood (Filho et al., 1993). Similar tendency was shown in CAT activity, PER level was not uniform and varied unclearly while GR activity demonstrated the opposite trend: it was less in pelagic fish than in animals belonging to other groups.

Wide antioxidant enzyme fluctuations were indicated within the ecological group especially suprabenthic/pelagic class in which enzyme activities differed in 2-7 fold. This was explained by the specificity of fish biology and life cycle. For instance the majority of examined species were tested in summer time, at the period of their reproduction, but *M. merlangus euxinus* spawns in winter. In examined period its physiological status was differed than the other fish species. Despite of intermediate antioxidant activities of CAT, PER and GR in blood of suprabenthic/pelagic fish (*S. smaris and S. acanthias*) we found high levels of the enzymatic activities in *M. merlangus euxinus* blood. So we could suggest that the relationship between antioxidant status and fish ecology modulated the abiotic factors and pollution which were agree with our previous data (Rudneva, 1997a, 1998a) and the results of other researchers (Ahmad et al., 2004; Amado et al., 2004a,b; Sole et al., 2009;).

Thus the activity of antioxidant enzymes in fish blood correlated with their swimming capacity. In our previous study we found that SOD and CAT activities in *T.meditrraneus* and *S. smaris* were significant higher than in the blood of slow swimming gobies, scorpion fish and flounder (Rudneva, 1997a). Other researchers reported also that CAT and SOD levels in blood of more active forms were higher as compared to more sluggish species (Filho et al., 1993). At the same time the changes in the sturgeon blood prooxidant-antioxidant status, as a consequence of adaptation to marine conditions, were not reflected in the liver and other tissues (Martinez-Alvarez et al., 2005).

No clear relationships between fish swimming capacity and non-enzymatic antioxidants could be established for the totality of fish muscle. The highest concentration of glutathione and fat-soluble vitamins was indicated in *M. barbatus ponticus* which was explained specificity of fish feeding and pigment formation mechanisms (Torrisen & Christiansen, 1995).

SOD, CAT and GR activities in fish muscle were higher in sluggish benthic forms as compared with suprabenthic and pelagic species. It agrees with the finding of Filho et al., (1993) who reported the decrease of antioxidant enzymes level in white muscle of active swimming fish species while in red muscle the opposite trends were demonstrated. High antioxidant enzymes level in red muscle of active fish was associated with high swimming activity and increase of metabolic rate which correlated with the increasing of ROS production and antioxidant defense induction (Zelinsky & Portner, 2000).

At present study high levels of antioxidant enzymes in the muscle of benthic fish could be correlated with pollution level in the bottom environment. We could propose that benthic and suprabenthic forms live in more contaminant environment because many xenobiotics accumulate in bottom sediments and low water layers. The long-term pollution impact on the fish could modify their defense system and change the adaptations to habitats. The other reason is linked with fish trophic level, feeding behavior and nutrition factors which also may affect antioxidant enzyme activities (Martinez-Alvarez et al., 2005; Sole et al., 2009). Benthic invertebrates (mollusks, crustacean and worms) and fish which are the preferable prey for these both fish groups might accumulate xenobiotics from the bottom sediments and transfer them via trophic nets to fish with the effect of concentration.

Liver of vertebrates exhibits a high metabolism and oxygen consumption and it is the main organ of xenobiotic detoxification. Fish liver displayed the highest levels of the key antioxidant enzymes SOD and CAT (Filho & Boveris, 1993; Filho et al., 1993; Rocha-e-Silva et al., 2004). SOD and CAT levels in liver appear to indicate that the most active species both teleosts and elasmobranches had greater enzyme activity compared with low mobile forms (Filho et al., 1993; Filho & Boveris, 1993). The higher activity of antioxidant enzymes in liver of active fish correlated with the higher oxygen consumption in fast swimming species and their high metabolic rate (Filho, 2007; Martinez-Alvarez et al., 2005). Animals with high metabolic rate exhibit the high rates of free radical production and cause the induction of antioxidant defense mechanisms (Zelinski & Portner, 2000).

In our studies we found high levels of non-enzymatic antioxidants in liver but correlations between fish ecological status and their contents were not shown. PER activity varied insignificantly, GR level was higher in active fish, CAT and SOD activities were greater in suprabenthic and suprabenthic/pelagic forms. At the same time our results were not shown the general trends of the increase of antioxidant enzymes activity in the liver of fast swimming fish as compared with sluggish animals which was documented by the other researchers (Filho et al., 1993; Filho & Boveris, 1993' Zelinsky & Portner, 2000). No clear relationships between fish ecological status and antioxidant enzyme defense in liver could be established for the species analyzed. We could proposed that our animals were caught in high polluted area (Sevastopol Bay) and the possible explanation could be that the benthic species induce their antioxidant defenses mechanisms as a response against environment pollution or accumulation of xenobiotics via food chains which was agreed with the data reported Sole et al., (2009). The induction of hepatic antioxidant activity evidences that fish from polluted sites are under oxidative stress and enzymes are involved in defense mechanism against peroxidative products (Amado et al., 2006 a,b). Pelagic and suprabenthic/pelagic forms have high levels of antioxidant enzymes as the adaptation mechanism against ROS production which was higher in active forms as compared to sluggish ones.

Fish trophic level, feeding behavior and nutrition factors also may affect biomarkers including antioxidants (Martinez-Alvarez et al., 2005; Sole et al., 2009). Our species were classed as pelagic plankton feeders (*T. mediterraneus*), omnivorous (*S. smaris*), carnivorous (*M. barbatus ponticus* and *N. melanostomus*) and predators (*S. acanthias, M. merlangus euxinus, P. maxima maeotica, S. porcus, P. flescuc lucsus*) (Table 1). No clear relationships between feeding groups and antioxidant enzyme activities in fish tissues could be detected also. However, predator fish species (*S. porcus*, *S. acanthias* and *M. merlangus euxinus*) demonstrated approximately similar enzymatic activity in the liver. It could be explained that benthic fish which are the preferable prey for such a group might accumulate xenobiotics from the bottom sediments and transfer them via trophic nets to fish with the effect of concentration. Hepatic antioxidant enzyme activity in pelagic plankton feeder *T. mediterraneus* and omnivorous suprabenthic/pelagic *S. smaris* was approximately similar with the exception of SOD which was explained the similarity of their high metabolic rate and diet consumption. Such a tendency was documented for hepatic GST activity in examined Black Sea fish species (Rudneva et al., 2010; 2011).

At the same time the information about relationships between fish feeding behavior and antioxidant defense is not uniform. SOD activity in liver showed no correlation with fish feeding behavior (Filho et al., 1993). In contrast, other authors documented the links between feeding and SOD activity level where high SOD value was found in herbivorous species as compared with carnivorous ones (Martinez-Alvarez et al., 2005). In contrast, taking into account that the most of sluggish fish species belonging to benthic and suprabenthic groups we could propose that they live in more contaminant environment because many pollutants accumulate in bottom sediments and low water layers. Thus, hard pressing of chemicals induces hepatic antioxidant activity especially in benthic and suprabenthic forms.

Obtained results demonstrated that the various characteristics of fish biology are reflected on antioxidants level in tissues. We could conclude that the complex of specific phylogenic, physiological and ecological features of fish species may modify antioxidant status and it is important to understand for development of monitoring programs. As we see antioxidant defense biomarkers of benthic forms are more convenient for monitoring studies, but benthic fish species are differed each from other, suprabenthic forms were more homogeneous as compared with suprabenthic/pelagic and benthic fish species. However, among suprabenthic species we found the forms with different swimming capacity and type of feeding. Present study indicates that the complex of biotic and abiotic factors including anthropogenic impact may be attributed to antioxidant defense in fish.

Conclusion

Thus the study of antioxidant status in Black Sea teleost and elasmobranch fish species allowed us to identify differences between them. We could suggest that the phylogenetic position and ecological status are reflected on the antioxidant defense in fish.

On the other side the interspecies variations of examined Black Sea fish species may be the result of the different organism sensitivity to chemical pollution. The tissue specific damage corresponded to the differences in the AOA potentials of the species for their adaptation to environmental stress (Ahmad et al., 2004). Previously we described the high anthropogenic impact on Sevastopol Bay (Black Sea, Ukraine) and its negative consequences on fish health (Rudneva & Petzold-Bradley, 2001; Rudneva et al., 2005; 2008; 2010a). The interspecies variations AOA defense may reflect the specific adaptations to the oxidative stress and protective mechanisms against ROS damage. Thus, the analysis of antioxidants level in different fish species taking into account their phylogenic position, specific features of ecology and physiology is important for the understanding the key ways of evolution of aerobic life, development of fish abilities to protect against pollutants and keep their life and biodiversity in the impact environments caused oxidative stress and damage animals health.

References

Ahmad I., Pacheco M., Santos M.A. *(2004). Enzymatic and nonenzymatic antioxidants as an adaptation to phagocyte-induced damage in Anguilla anguilla L. following in situ harbor water exposure.* Ecotoxicol. and Environ. Safety. V. 57. P. 290-302.

Amado L.L., da Rosa C.E., Leite A.M., Moraes L., Pires W.V., Pinho G.L.L., Martins C.M.G., Robaldo R.B., Nery L.E.M., Monserrat J.M., Bianchini A., Martinez P.E., Geracitano L.A. *(2006a). Biomarkers in croakers Micropogonias furnieri (Teleostei: Sciaenidae) from polluted and non-polluted areas from Patos Lagoon estuart (Southern Brazil): evidences of genotoxic and immunological effects.* Mar. Pollut. Bull.V. 52. P.199-206.

Amado L.L., Robaldo, R.B., Geracitano, L.A. Monserrat, J.M., Bianchini, A. (*2006b). Biomarkers of exposure and effect in the Brazilian flounder Paralichthys orbignyanus (Teleostei: Paralichthyidae) from the Patos Lagoon estuary (Southern Brazil): Evidences of genotoxic and immunological effects.* Mar. Pollut. Bull. V.52. P. 207-213.

Blazer V.S. *(1992). Nutrition and disease resistance in fish.* Ann. Rev. Fish Diseases.V. 2, P. 309-323.

Burlakova E.B. *(2005). Bioantioxidants: yesterday, today, tomorrow. In: Chemical and Biological kynetics. New approach.* Moscow. Chemistry Publ. V. 2. P. 10-45. (*in Russian*)

Dabrowski K. & Ciereszko A. *(2001). Ascorbic acid and reproduction in fish: endocrine regulation and gamete quality*. Aquaculture Res. V. 32. P. 623-638.

Dafre A.L. & Reischl E. *(1990). High hemoglobin mixed disulfide content in hemolysates from stressed shark.* Comp. Biochem. Physiol. V. 96B. P. 215-219.

Filho W.D.*(1996). Fish antioxidant defenses: a comparative approach.* Braz. J. Med. Biol. Res. V. 29 (12). P. 1736- 1742.

Filho W.D. *(2007). Reactive oxygen species, antioxidants and fish mitochondria.* Fishery in Bioscience. V.12. P. 1229-1237.

Filho D.W., Giulivit C., Boveris A.*(1993). Antioxidant defenses in marine fish. – I. Teleosts.* Comp. Biochem. Physiol. V. 106C (2). P. 409-413.

Filho D.W., Boveris A.*(1993). Antioxidant defenses in marine fish. – II. Elasmobranchs.* Comp. Biochem. Physiol. V. 106C (2). P. 415-418.

Filipovich U.B., Egorova E.A., Sevast'yanova G.A *(1975). Manual Book of General Biochemistry.* Moscow. Prosveshenie Publ. . P. 200 – 203 (*in Russian*).

Folomeev V.F. *(1981). Photocolorimtric micromethod of SH-groups content determination.* Laboratornoe Delo. V. V. P. 33-35. (*in Russian*)

Goldberg D.M. & Sparner R.J. *(1987). Glutathione reductase. In: Bergmeyer, H.U., Bergmeyer, J., Grab, M. (Eds) Methods of Enzymatic Analysis.* V. III. Verlag Chemic, Weinheim. 258-265.

Halafian A.A. *(2006). Statistica.* Moscow, Binom Publ. 512 pp.

Hardy R.W. *(2001). Nutritional deficiency in commercial aquaculture: likelihood, onset and identification.* In: Nutrition and Fish Health (Ed. Ch.Lim & Webster C.D.). N.Y., Oxford. P. 131-148.

Hussein G., Sankawa U., Goto H., Matsumoto K., Watanabe H. *(2006). Astaxantin, a carotenoid with potential human health and nutrition.* J. Nat Prod. V. 69. P. 443-449.

Karnauhov V.N. & Fedorov G.G. *(1982). Methods of carotenoids and vitamin A determination in animal tissues.* Methods recommendations. Pushino Publ. P..15-18. *(in Russian)*

Lali S.P. &Lewis-McCrea L.M. (2007). *Role of nutrients in skeletal metabolism and pathology in fish – an overview.* Aquaculture. V. 267. P. 3-19.

Lesser M. P. *(2006). Oxidative stress in marine environmnets: biochemistry and physiologycal ecology.* Ann. Rev. Physiol. V. 68. P. 253-278.

Li S.-D., Su Y.-D., Li M., Zou Ch.-G. *(2006). Hemin-mediated hemolisis in erythrocytes: effects of ascorbic acid and glutathione.* Acta Biochem. Biophys. Sinica. 38(1.) P. 63-69.

Litvin F.F. *(1981). Laboratory Manual of Physicochemical Methods in Biology.* Moscow: Moscow State University Publ. P. 86-87. (*in Russian*)

Livingstone D.R. *(2001). Contaminant-stimulated reactive oxygen species production and oxidative damage in aquatic organisms.* Mar. Pollut. Bull. V. 42. P. 656-665.

Lowry O.H., Rosebrough N.J. Farr, A.L. Randall R.J. *(1951). Protein measurement with the Folin-phenol reagent.* J.Biol.Chem. V.193. P. 165-175.

Martinez-Alvarez R.M., Morales A.E., Sanz A.*(2005). Antioxidant defenses in fish: biotic and abiotic factors.* Reviews in Fish Biology and fisheries. V. 15. P. 75-88.

Nishikimi M., Rao N.A., Yagik K. *(1972).The occurance of superoxide anion in the reaction of reduced phenazine.* Biochem. Biophys. Res. Commun. V. 46 (2.). P. 849 -854.

Papp L.V., Lu J., Holmgren A., Khanna K.K. *(2007). From selenium to selenoproteins: synthesis, identify and their role in human health.* Antioxidants and Redox Signaling. V. 9 (7). P. 775-793.

Peters L.D., Livingstone D.R. *(1996). Antioxidant enzyme activities in embryonic and early larval development of turbo.* J. Fish Biol. V. 49. P. 986-997.

Putilina F.E. *(1982). Methods of Biochemistry.* Leningrad: Leningrad Unuiversity. P. 183-186. (*in Russian*)

Rocha-e-Silva, T.A.A., Rossa, M.M., Rantin, F.T., Matsumura-Tundisi, T., Tundisi, J.G., Degterev, I.A.*(2004). Comparison of liver mixed-function oxygenase and antioxidant enzymes in vertebrates.* Comp. Biochem. Physiol. V. 137C. P. 155-165.

Rudneva I.I. *(1997a). Blood antioxidant system of Black Sea elasmobranch and teleosts.* Comp. Biochem. Physiol. V. 118C. P. 255-260.

Rudneva I.I. (1997b). *Ecological and phylogenetic peculiarities of antioxidant enzyme activities and antioxidant content in Black Sea teleosts and elasmobranch.* J. Evol. Biochem. Physiol. V. 33. P. 29-37. (*Saint-Petersburg, in Russian*)

Rudneva I.I. *(1998a). The biochemical effects of toxicants in developing eggs and larvae of Black Sea fish species.* Marine Environ. Res. V. 46. P. 499-500.

Rudneva I.I. *(1998b). Ecological and phylogenetic peculiarities of lipid composition and lipid peroxidation in teleosts and elasmobranch in Black Sea.* J. Evol. Biochem. Physiol. V. 34. P. 310-318. (*Saint-Petersburg, in Russian*)

Rudneva I.I. *(1998c). The response of blood and liver antioxidant system in two Black Sea fish species as biomarkers of pollution effects.* Extended Synopsis of International Symposium on Marine Pollution. Monaco, Oct. 5-9. 587-588.

Rudneva I.I. *(1999). Antioxidant system of Black Sea animals in early development* Comp. Biochem. Physiol. V. 122C (2) P. 265-271.

Rudneva I.I., Petzold-Bradley E. *(2001). Environmental and security challenges in the Black Sea region.* In: E. Petzold-Bradley, A. Carius and A. Vimce (Eds.), Environment Conflicts: Implications for Theory and Practice. Netherlands: Kluwer Academic Publishers, pp. 189–202.

Rudneva I.I., Shevchenko N.F., Zalevskaya I.N., Jerko N.V. *(2005) Biomonitoring of the coastal waters of Black Sea.* Water Resources. V.32 (2). P. 238-246.*(in Russian)*

Rudneva I.I., Skuratovskaya E.N., Omelchenko S.O., Zalevskaya I.N. *(2008). The application of fish blood biomarkers for the ecotoxicological evaluation of marine coastal waters.* Ecological Chemistry. V. 17 (2), P. 77 – 84. (*in Russian*)

Rudneva I.I., Kuzminova N.S., Skuratovskaya E.N., Kovyrshina T.B. (*2010a). Age composition and antioxidant enzyme activities in blood of Black Sea teleosts.* Comp. Biochem. Physiol. Vol.151C. P. 229–239.

Rudneva I.I., Kuzminova N.S., Skuratovskaya E.N. *(2010b).Glutathione-S-Transferase activity in tissues of Black Sea fish species.* Asian J. Exp. Biol. Sci. V. 1 (1). P. 141-150.

Rudneva I.I., Kuzminova N.S., Skuratovskaya E.N., Kovyrshina T.B.*(2011). Comparative study of Glutathione-S-Transferase activity in tissues of Black Sea teleosts.* Int. J. Science and Nature. V. 1 (1). In press.

Sole M., Rodriguez S., Papiol V., Maynou F., Cartes J.E. *(2009). Xenobiotic metabolism markers in marine fish with different trophic strategies and their relationship to ecological variables.* Comp. Biochem. Physiol.V. 149C(1). P. 83-89.

Svetovidov A.N. *(1964). Black Sea Fishes.* Leninhgrad. Nauka Publ. 350 pp. (*in Russian*).

Torrisen O.J. & Christiansen R. (*1995). Requirements of carotenoids in fish diets.* J. Appl. Ichthyol. V. 11. P. 225-230.

Vladimirov Ju. A. & Proskurina E.V. *(2007). Lectures of Medical Biophysics.* Moscow University Publ. «Academkniga». 432 pp. (*in Russian*)

Winston G. *(1990). Physicochemical basis for free radical formation in cells: production and defenses.* In: Stress responses in plants: adaptation and acclimation mechanisms P. 57-86.

Winston G.W. *(1991). Oxidants and antioxidants in aquatic organisms*. Comp. Biochem. Physiol. V. 100 C (1-2). P. 173-176.

Winston, G.W. & Di Giulio, R.T. *(1991). Prooxidant and antioxidant mechanisms in aquatic organisms. Aquatic Toxicol.* V. 19. P. 137-161.

Zelinski S. & Portner, H-O.*(2000). Oxidative stress and antioxidative defense in cephalopods: a function of metabolic rate or age*? Comp. Biochem. Physiol. V.125B. P. 147-160.

In: Fish Ecology
Editor: Sean P. Dempsey

ISBN 978-1-61324-282-7

Chapter 3

DOES HABITAT AFFECT THE GENOMIC GC CONTENT? A LESSON FROM TELEOSTEAN FISH: A MINI REVIEW

Ankita Chaurasia[1], Erminia Uliano[2], Luisa Bernà[1], Claudio Agnisola[2] and Giuseppe D'Onofrio[1,]*

[1]Laboratory of Animal Evolution and Physiology. Dept. Genome Evolution and Organisation. Stazione Zoologica Anton Dohrn. Villa Comunale – 80121 Napoli, Italy
[2]Department of Biological Sciences. University of Naples Federico II.Via Mezzocannone, 8 -80134 Napoli,Italy

ABSTRACT

The nature of the forces driving the genomic GC content of both prokaryotes and eukaryotes is matter of debate. Latest results favor selection as the main factor shaping base composition in bacteria. In vertebrates the subject is still under discussion.

Focusing on teleostean fish, the mass specific routine metabolic rate temperature-corrected using the Boltzmann's factor (MR) and base composition of genomes (GC%) were re-examined and related with their major habitats: polar, temperate, sub-tropical, tropical and deep-water. Fish of the polar habitat showed the highest MR. The MR of temperate fish was significantly higher than that of tropical one, which showed the lowest average value. Regarding GC%, polar and temperate fish both showed significantly higher values than both sub-tropical and tropical fish. Plotting MR and GC% a significant correlation was found between the two variables.

Different methylation levels characterized the genomes of fish living in different habitats. More precisely, the amount 5-methylcytosine (5mC) decreases from polar to tropical fish. Considering the positive correlation between CpG and GC%, as well as the temperature dependence of the 5mC deamination process, low genomic GC% in fish living in warm habitats could be the result of a CpG shortage.

The frequencies of CpG as well as that of the derivative doublets, that is TpA and CpA, were checked in the intronic sequences of the available teleostean genomes

* Corresponding author, donofrio@szn.it

completely sequenced, namely *D. rerio*, *O. latipes*, *G. aculeatus*, *T. nigroviridis* and *T. rubripes.* The analysis of doublets further supported a link between environment, metabolic adaptation and genome base composition.

In this frame, also taking into account that MR turned out to be not significantly different between fish living in tropical and deep-water habitats, we suggest that the level of environmental O_2 content (dictated by the Henry's law) could be a source of variability that directly affects MR adaptation and hence the base composition of fish genomes.

Neutral and selective hypotheses, proposed to explain the base composition evolution at the genome level, namely mutational bias, biased gene conversion (BGC), DNA breakpoints distribution, thermal stability and metabolic rate, were discussed in the light of current data. Negative (purifying) selection is proposed for the genome variation of teleostean fish among habitats.

Keywords: genome evolution; metabolic rate; CpG; methylation; 5-methylcytosine; BGC; thermal stability

Abbreviations

5mC	5-methylcytosine
AT	molar ratio of adenine plus thymine
BGC	biased gene conversion
bp	base pair
bpi	base pair in introns
CpA	cytosine nucleotide occurring next to adenine nucleotide
CpG	cytosine nucleotide occurring next to guanine nucleotide
GC	molar ratio of guanine plus cytosine;
GC3	molar ratio of guanine plus cytosine at third codon positions;
GCi	molar ratio of guanine plus cytosine in intron
HACNS	human-accelerated conserved non-coding sequences
HAR	human-accelerated region
HPLC	high-performance liquid chromatography
MLB	metabolic level boundaries
MR	mass specific routine metabolic rate, temperature-corrected using the Boltzmann's factor
SNP	single nucleotide polymorphism
T°	temperature
TpA	thymine nucleotide occurring next to adenine nucleotide

Introduction

The genome base composition, generally defined as the genomic GC content, i.e. the molar ratio of guanine plus cytosine, widely varies among genomes, especially in bacteria where it ranges from less that 20% to more than 70% GC, as first described using the density gradient centrifugation analysis by Sueoka [1]. From that time on, several hypotheses have been proposed to explain the nature of the forces driving the GC content variability among genomes. A still open debate in which internal forces are opposed to the external ones. The former are mainly based on stochastic events arising during intracellular processes, such as DNA duplication, repair, recombination. The latter takes into account the role of adaptive processes resulting from the interaction of the organism with the surrounding environment. Thus, the neutralist-selectionist debate remains the "heart of the problem" in the field of molecular evolution, and recently was further fueled by the corrosive question "are there any neutral sites in the genomes of bacteria?" [2].

In the frame of the neutral theory, mutational bias hypothesis [3-5] was first proposed to explain the great variation of the genomic GC content in bacteria, and later on extended to high vertebrates [6]. The rational of the hypothesis was based on the observation that GC variation takes place mainly at sites under weak selective pressure, i.e. the third codon position (GC3) and non-coding regions, considered of little adaptive relevance. However, further analysis of the mutation pattern in *E. coli* [7] and in human [8,9], showed an excess of GC ⇒AT mutations, hardly explaining the preservation of the current genomic GC content in bacteria [10] and the compositional heterogeneity of the human genome characterized by the presence of the GC-rich isochores [11].

In the same frame, the biased gene conversion hypothesis (BGC), essentially based on a synergy between recombination events and biased DNA repair [12-14], was first proposed to explain the formation of GC-rich isochores in mammals [12], and afterward extended to prokaryotes, since the finding of a GC biased repair system in those organisms too [15].

The weak point of the BGC hypothesis is that, in spite of the fact that the compositional structure of the mammalian genomes (i.e. isochore organization) appears to be well conserved [16-18], hot-spot recombination sites were reported to be not phylogenetically conserved among mammals, even in close related species such as chimpanzee and human [19]. Moreover, the size of the elements known to evolve according to the BGC, namely HARs and HACNSs, is within the range of 200-1000 bp [20], an order of magnitude far from the mega-bases of the isochores [21].

In bacteria, very recent analysis of the mutation pattern, analyzed in about 150 genomes, showed the existence of a universal AT pressure [22]. The authors suggested that the natural selection and/or a natural selection-like process, i.e. BGC, could affect the nucleotide frequencies [22]. However, detailed analysis of bacterial genomes, carried out to assess the single nucleotide polymorphism (SNP) at synonymous positions, failed to give support to the BGC hypothesis [23]. An excess of GC ⇒AT polymorphism at synonymous positions was observed, indeed, in genomes with no evidence for recombination, living open the question: "why this is so" [23].

In the frame of the selectionist point of view, several environmental factors, showing a correlation with the DNA base composition, have been reported in favor of an adaptive hypothesis in bacteria: competition for metabolic resources [24], endosymbiosis [25], environments/habitats [26], growth temperature [27], and environmental oxygen availability (i.e., aerobiosis vs. anaerobiosis, 28, 29]. The last three papers stressed the effect on the genome GC content of two main adaptive environmental factors: temperature and metabolic rate. These factors were also claimed to affect vertebrate genomes and largely discussed in the thermal stability hypothesis [21] and in the metabolic rate hypothesis [30-32].

According to the thermodynamic stability hypothesis, it is expected that the increment of environmental or body temperature produces a GC increment, stabilizing DNA, RNA and proteins [11, for a review]. The hypothesis was based on two main points: i) the fact that

DNA and RNA structure are stabilized by the triple hydrogen bonds of the GC pair; and ii) on the observation that hydrophobicity, well known to play a stabilizing role of the protein structure [33], was significantly increasing in proteins encoded by genes localized in GC-rich isochores [34], a point unfortunately misapprehended by Eyre-Walker and Hurst [35], because of wrong statistical assumptions, as highlighted by Clay and colleagues [36].

The metabolic rate hypothesis was based on the observation that two DNA features, bendability [30] and nucleosome formation potential [31], were both significantly correlated with the GC content. More precisely, an increment of the DNA bendability [30] paralleled by a decrement of the nucleosome formation potential [31] favor the GC-rich genes, that are reported to be characterized by a high transcriptional activity [37,38]. The analysis of human functional classes further supported the hypothesis. Indeed, the GC3 level of the genes involved in cellular metabolism was significantly higher than those involved in information storage and processing [39].

With the aim to shed light on the selectionist/neutralist debate about the forces shaping the variability of the GC content among genomes, we focused on teleostean fish. The rational of the choice grounded on the consideration that a marine organism, differently from terrestrial one, lives in an environment where the available oxygen, dictated by the Henry's law, is a limiting factor. Moreover, this eco-physiological condition allows disentangling the oxygen consumption from the environmental temperature. In particular, the data on metabolic rate and genomic GC content were critically reanalyzed on the light of: i) the different levels of 5-methylcytosine (5mC) observed in the genomes of fish living in different habitats [40]; and ii) the effect of the temperature on the spontaneous deamination of 5mC [41].

The metabolic rate and the genome base composition in teleostean fish.

Metabolic processes allow organisms to transform the energy and material taken from the environment into their structures and functions [42]. Metabolism determines the energetic demands the organisms place on their environment. On the other hand, ecological factors, affecting both physiology and selective pressure, will influence the organismic performance and homeostasis, and hence its energetic demands. This reciprocal interaction between environment and metabolism is likely to represent a major determinant of life-history evolution, and affects numerous biological properties and rates, including rate of DNA evolution and DNA diversity [43]. Organismic metabolic rate is mainly related with body size and temperature [42]. Mass specific metabolic rate varies with body size elevated to a factor that approximates a quarter-power [44-46]. At present it is still questioned if this represents a universal scaling law. The debate is mainly dealing with the meaning of the species-related variability in both the scaling coefficient (a) and the scaling exponent (b) of the allometric equation ($R = aM^b$, where R is the resting metabolic rate and M is the body mass) [47-52]. Recently a curvilinear rather than linear relationship between LogR and LogM has been proposed [53]. A metabolic level boundaries (MLB) hypothesis stressing on the boundary constrains that limit the scaling of metabolic rate has also been proposed [54] and applied to teleost fish [52] suggesting that lifestyle, swimming mode and temperature affect the intraspecific scaling of resting metabolic rate.

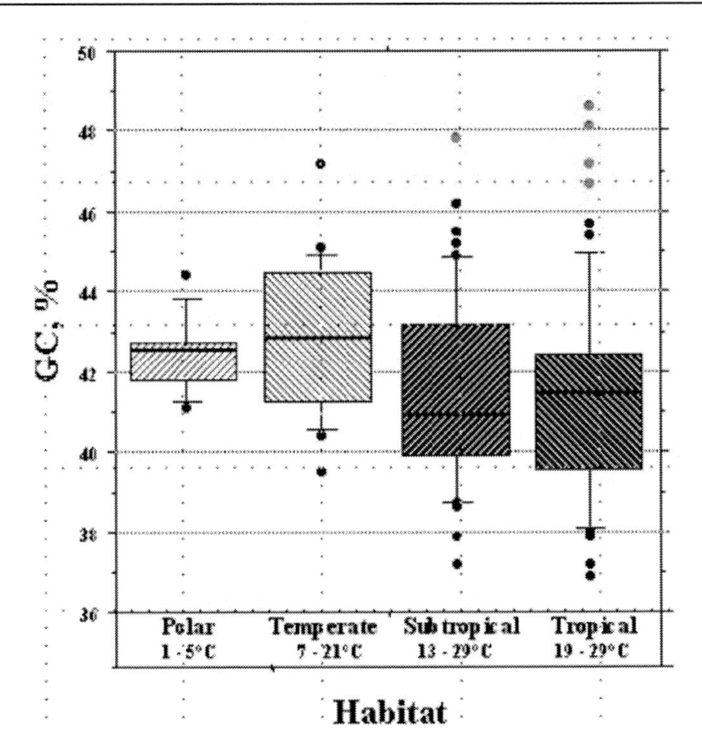

Figure 2.Box plot of genomic GC% distributions within each habitat group, sorted according to the temperature range. Outliers are shown (red dots). Modified from [55].

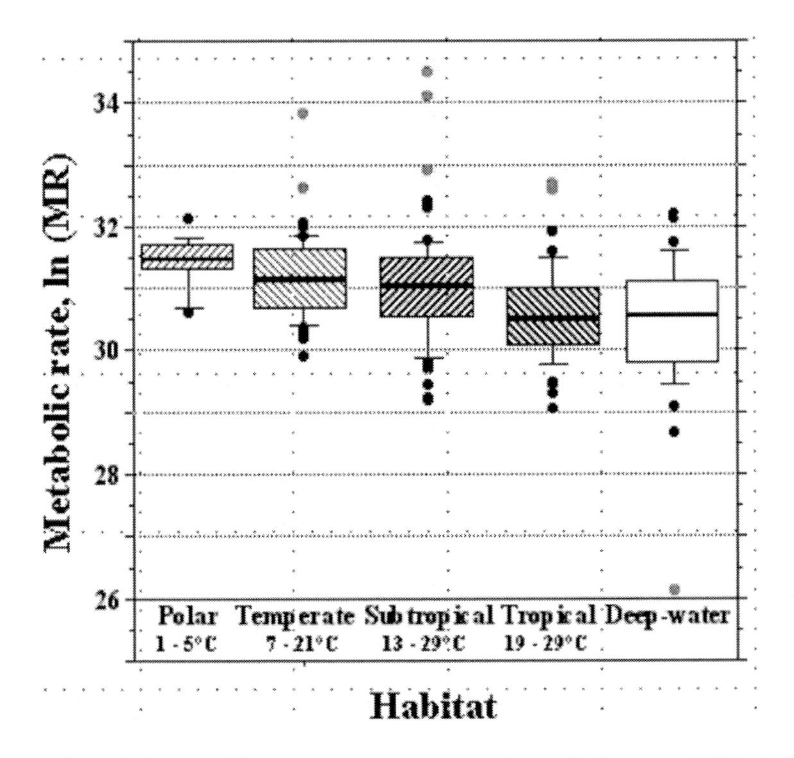

Figure 1. Box plot of the specific, temperature corrected metabolic rate (MR) log-normalized distributions within each habitat group Boxes are sorted according to the increasing average temperature, except for deep-water. Outliers are shown (red dots). Modified from [55].

Temperature is likely to affect metabolism mainly via its effects on the rates of biochemical reactions. The temperature dependence can be modeled in terms of Q10 [51] or using the Boltzmann's factor (or Van't Hoff-Arrhenius relation, $e^{-E/kT}$) [52,55]. There are several reasons to prefer the Boltzmann's factor to Q10 [56]. Q10 is assumed to be temperature independent, while the temperature dependence of biological processes usually is not purely exponential [57]. Moreover, the Q10 value can vary in a wide range and has been shown to be species specific [58]. Indeed, an error up to 15% may be introduced evaluating the temperature dependence of metabolism using Q10 values [59]. To the aim of the present paper, values of oxygen consumption were corrected using the Boltzmann's factor. The temperature corrected mass specific resting metabolic rate (MR) can be utilized to compare metabolism in different species independently from differences in body size and temperature [52,55], and correlate metabolism with morpho-functional characteristics at molecular level, such as genome base composition, in order to infer on their possible adaptive significance.

Data about MR were collected for a large set of fish (>200) from fishbase.org, and the data set was divided in five habitat groups, namely polar, temperate, subtropical, tropical and deep-water (Figure 1). Teleostean fish of the polar group, characterized by a small range of living temperature (1-5°C), showed the highest average MR value, significantly higher ($p< 2.75 \times 10^{-2}$) even than that of fish living in temperate habitat (7-21°C). The latter showed an average metabolic rate slightly, but not significantly higher, that those living in the subtropical habitat (13-29°C). On the contrary, fish living in the tropical habitat (19-29°C) showed an average MR significantly lower than those living in the temperate one ($p<1.0 \times 10^{-4}$), as well as lower than those living in the subtropical habitat ($p<1.1 \times 10^{-3}$). Interestingly, the average MR of deep-water fish was not significantly different from that of tropical fish.

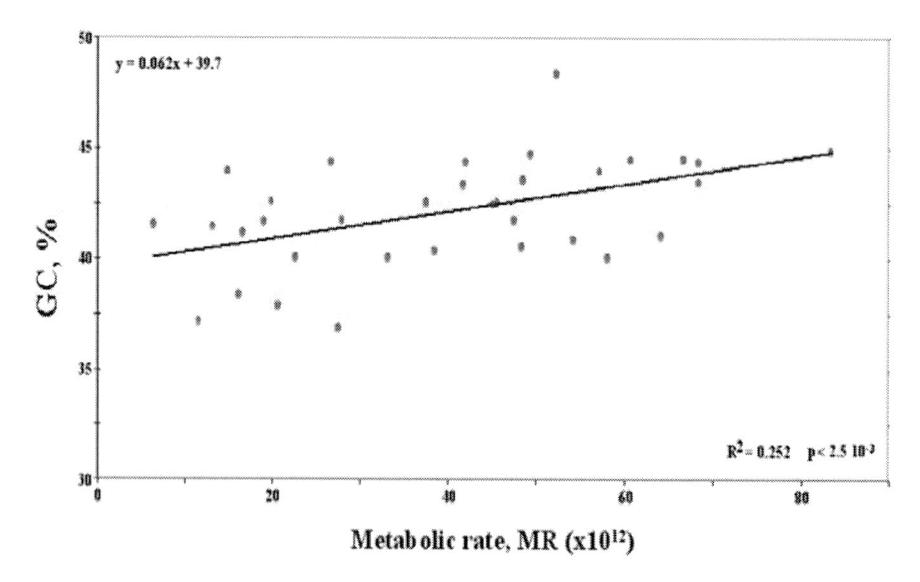

Figure 3. Plot of the specific, temperature corrected metabolic rate (MR), against the average genomic GC content (GC%). The equation of the linear regression and the correlation coefficient (R) and the statistical significance of the regression (*p*-value) are reported. Modified from [55].

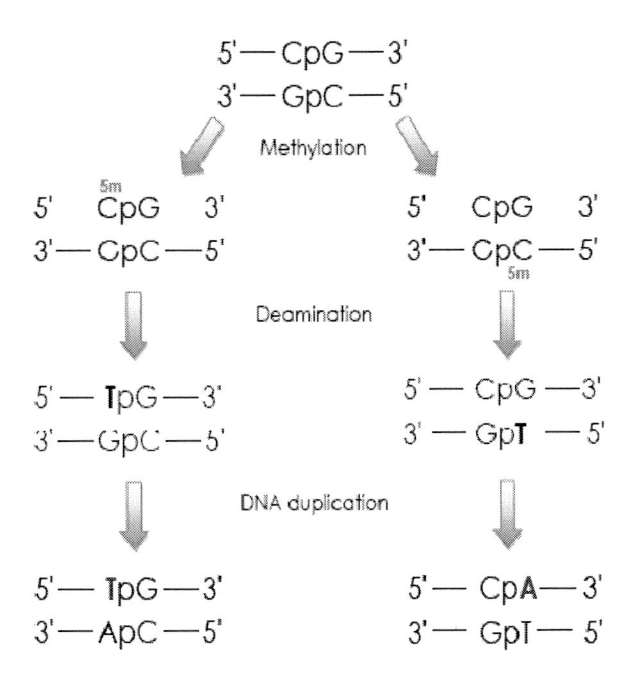

Figure 4. Scheme of the methylation/deamination/ process of CpG giving rise to TpG and CpA doublets.

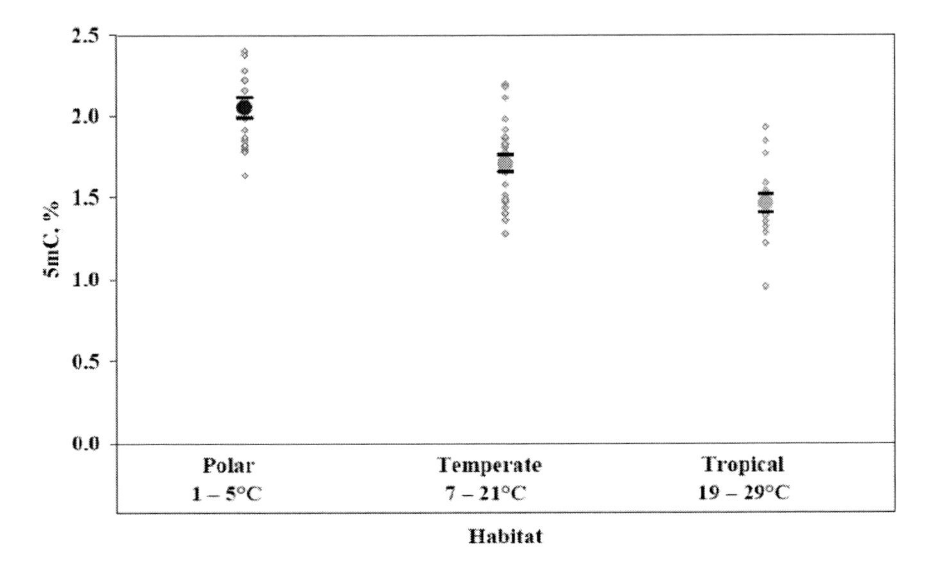

Figure 5.The cytosine methylation levels (5mC) in the genomes of fish living in different habitats.Modified from [40].

The data reported in Figure 1 were of great interest because: i) contrary to the expectation, the average temperature of the habitats, without considering the deep-water habitat, was negatively correlated with the MR; and ii) deep-water and tropical fish, in spite of a different living temperature, showed a MR not significantly different. Moreover, the analysis of a sub-set of fish belonging to the class of Perciformes led to the same conclusions [55] supporting the point of view that the MR was independent from the phylogenetic relationship among species [56]. A logical conclusion from the above considerations is that temperature does not affect directly the MR of fish living in different habitats; more reasonably, the main factor is the amount of free O_2 available in the environment, i.e. dissolved in the water, according to the Henry's law.

In order to answer to the open question of the present review, the genomic GC content of fish was retrieved from current literature [60]. In that paper, data about the genome base composition of more than 150 teleostean fish were obtained by the analytical centrifugation technique and the known correlation between GC content and buoyant densities [61]. The data of the genomic GC content of fish were analyzed according to the different habitats. The box-plot analysis clearly showed that the genomic GC values were not randomly distributed, but related to the habitats (Figure 2). Indeed, the average genomic GC content of fish living in both polar and temperate habitat was significantly higher than that of fish living in subtropical and tropical habitat (*p*-values ranging from 5.5×10^{-2} to 5.0×10^{-3}). Interestingly, both MR and genomic GC content showed a negative trend towards the average temperature of the habitats. Crossing MR and GC dataset, data about 34 species were available. Interestingly, a significant positive correlation was found between the two variables (Figure 3).

The Role of the CpG Doublet

It is well known, since the paper of Josse and colleagues [62] that CpG is the most underrepresented doublet according to the frequency of all possible DNA doublets. The reason of the so-called CpG-shortage was found in the high rate of mutation of this doublet due to the methylation process followed by a deamination of the 5mC [63]. Indeed, while the deamination of unmethylated cytosine turn out to uracil, that is removed by enzymatic reaction [64], the deamination of the 5mC turn out to thymine [65], that is fixed at DNA level because it cannot be repaired. Therefore, methylated CpG have been reported in the literature as highly responsible of the GC $\Rightarrow$AT compositional changes [66]. As shown in Figure 4, the methylation/deamination/ process of cytosine, affecting the CpG doublets present in both DNA strands, is traceable through the frequencies of the derivative doublets. Indeed, the loss of 5mCpG produces the gain of TpG and CpA doublets [67].

Comparative genome analyses led to the observation that, on average the genomes of fishes and amphibians the methylation level is twice than in mammals, birds and reptiles [68, 69]. Moreover, the methylation level of fish genomes was reported to be different in fish living in different habitats (Figure 5), more precisely decreasing from polar to tropical [40]. Taking into account that the deamination process affecting the 5mC was reported to be temperature dependent [41], the role played by CpG, TpG and CpA doublets in shaping the GC content of fish genomes living at different temperatures was checked. To this aim the non-coding sequences, more precisely the intronic sequences, of the following organisms, for which the full genomic sequence is available, were analyzed: stickleback (*G. aculeatus,* 44.5% GC), fugu (*T. rubripes,* 44.0% GC), zebrafish (*D. rerio,* 36.9% GC), medaka (*O. latipes,* 40.1% GC) and pufferfish (*T. nigroviridis,* 45.6% GC). Intron analysis allows, through the gene orthology, to compare corresponding DNA sequences in different genomes. Moreover, the methylation/deamination is not a region specific process, indeed affecting indistinctly all non-coding DNA stretches.

The box-plots of CpG and TpG + CpA (showing minimum and maximum values, the lower and the upper quartile, the median of the distribution, but, for graphical reasons, not the outliers) were arranged according to the increasing living temperature of each organism (Figure 6; top and bottom panel, respectively). Regarding CpG, both zebrafish and medaka showed a frequency that was two times lower than that of other fish (Figure 6; top panel). The result was not surprising, since in all genomes the frequency of the CpG doublet was highly correlated with the genomic GC% (Table 1), as expected according to previous reports [70,71]. Regarding TpG + CpA, although zebrafish and medaka showed again the lowest values, small differences were observed among genomes (Figure 6; bottom panel). The result showed in Figure 6 could be explained by the occurrence, especially in the organisms showing very low CpG frequencies, of a high negative correlation between CpG and TpG or CpA. The correlation between CpG and the derivative doublets was checked at both inter-and intra-genomic level. In the first case the correlation turned out to be not significant. Indeed, the *p*-values of the Spearman rank correlation test of CpG *vs* TpG and CpG *vs* CpA were $p<0.16$ and $p<0.96$, respectively. The intra-genomic correlations also showed very poor correlation coefficients, especially low in zebrafish and medaka (Table 1, top). Moreover, plotting the CpG frequency against that of TpG + CpA, the negative correlation was higher in fish living at lower than at higher environmental temperature (Figure 7). It is well known that transposable and repetitive elements, largely occurring in intronic sequences and affecting their evolution [72], can represent more than 10% of the genome, as the case of the smallest vertebrate genome, i.e. *T. nigroviridis* [73]. By reiterating the analysis after removing interspersed repeats and low complexity DNA sequences by RepeatMasker, the overall conclusion kept on unchanged (Table 1, bottom), although several correlations showed increased R values.

Due to the very different numbers of intronic sequences available for each fish genome (ranging from the $10^4$of medaka to the $3.8x10^4$ of fugu), and in order to avoid any bias because of different sets of genes, the analyses were restricted to sets of orthologous pairs of introns. It is worth to stress that orthology was based on the analysis of coding sequences and extended to the corresponding intronic sequences. The ending results confirmed previous observations. For example, in the set of orthologous intron pairs: i) zebrafish *vs.* stickleback: the correlation coefficients (R) of the intragenome regression CpG *vs.* TpA+CpA were 0.133 and 0.314, respectively; ii) medaka *vs.* stickleback: the R values were 0.004 and 0.289, respectively; iii) zebrafish *vs.* fugu: the values were 0.133 and 0.316, respectively; and iv) medaka *vs.* fugu: the values were 0.013 and 0.312, respectively. Again by removing repetitive elements, intragenomic correlations between doublets, using orthologous sets, turned out to be barely affected (data not shown). In all pairwise comparisons of orthologous intron sequences, zebrafish and medaka, showed low frequencies for the CpG doublet, as well as also low correlation with the derivative doublets of the methylation/deamination process, i.e. TpA and CpA. In this frame, it is worth to bring to mind that at least in zebrafish, differently from other organisms, no correlation was found between SNPs and CpG [74].

In short, the frequencies of the CpG doublets turned out to be very poorly affected by the methylation/deamination process. Is it worth to recall that increments of the GC content, and hence the frequency of the CpG doublets, influence the histone acetylation level and consequently the chromatin structure [75], a result in agreement with the observed correlation between metabolic rate and genomic GC content [55].

Table 1. Intragenomic Correlation coefficient (R) before and after Repeat Masker[a]

Species	GC vs CpG	CpG *vs* TpG	CpG *vs* CpA
G. aculeatus	0.773	0.188	0.223
T. rubripes	0.803	0.166	0.213
D. rerio	0.686	0.006	0.076
O. latipes	0.741	0.010	0.037
T. nigrovoridis	0.837	0.118	0.241
G. aculeatus	*0.781*	*0.238*	*0.282*
T. rubripes	*0.820*	*0.227*	*0.261*
D. rerio	*0.693*	*0.077*	*0.044*
O. latipes	*0.746*	*0.007*	*0.056*
T. nigrovoridis	*0.843*	*0.197*	*0.324*

[a]Values in italic.

Materials and Methods

Genome Sequences

Coding sequences of the finished genome assembly for all five fish namely *D. rerio* (Assembly: Zv7, Apr 2007, Ensembl Release: 48.7b), *O. latipes* (Assembly: HdrR, Oct 2005, Ensembl Release 48.1d), *G. aculeatus* (Assembly: BROAD S1, Feb 2006, Ensembl Release 48.1e), *T. nigroviridis* (Assembly: TETRAODON 7, Apr 2003, Ensembl Release 48.1j), and *T. rubripes* (Assembly: FUGU 4.0, Jun 2005, Ensembl Release 48.4h) were retrieved from the ENSEMBL (http://ftp.ensembl.org).

Non-Coding sequences were retrieved from UCSC Genome browser (http://genome.ucsc.edu/), for *D. rerio* (Assembly: Apr 2007, Zv7/danRer5), *O. latipes* (Assembly: Oct 2005, NIG/UT, MEDAKA 1/ oryLat2), *G. aculeatus* (Assembly: Feb 2006, BROAD/gas Acu1), *T. nigroviridis* (Assembly: Feb 2004, Genoscope 7.0/tetNig1), and *T. rubripes* (Assembly: Oct 2004 (JGI 4.2/ fr2). In each genome the number of full length genes (i.e. CDS+introns) was: *D. rerio* 24507, *O. latipes* 10230, *G. aculeatus* 23191, *T. nigroviridis* 18384, *T. rubripes* 38200. Sequences containing characters indicating ambiguity in identification of certain bases, i.e. N, were discarded.

RepeatMasker (Version 3.1.9, http://www.repeatmasker.org) was used to mask the interspersed repeats and low complexity DNA sequences. (A.F.A. Smit, R. Hubley and P. Green). The overall number of sequences was basically unaffected by the masking process. The final data set analyzed consisted of *D. rerio* (24475), *O. latipes* (10212), *G. aculeatus* (23156), *T. nigroviridis* (18317), and *T. rubripes* (18317).

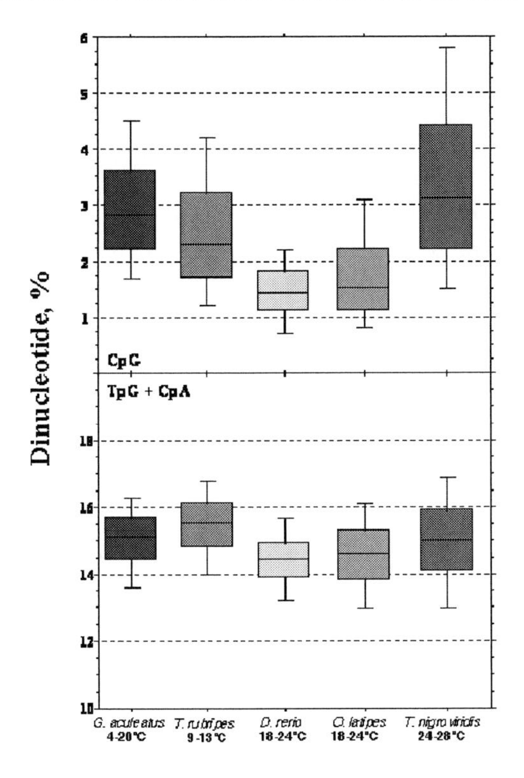

Figure 6. Box plot of the CpG (top panel) and TpA+CpA (bottom panel) frequencies in the intronic sequences of fish living at different temperatures.

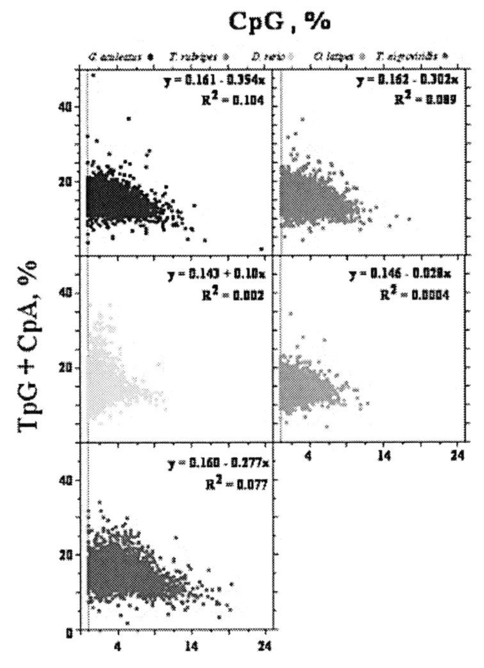

Figure 7. The plots show the correlations between CpG and TpA+CpA in the intronic sequences of fish living at different temperatures. The equations of the regression lines, as well as the correlation coefficients (R) are reported.

Orthologs

Orthologous gene pairs were identified using a Perl script, which performs reciprocal Blastp [90] and selects the Best Reciprocal Hits (BRH), hence identifies the true orthologs by avoiding the consideration of paralogous sequences.

All the pair wise comparative analyses on non-coding sequences were performed between the orthologous gene pairs obtained from the above method, for all the possible ten combinations among the five fish namely, *D. rerio -O. latipes* (4765); *D. rerio -G. aculeatus* (9516); *D. rerio -T. nigroviridis* (8848); *D. rerio -T. rubripes* (9792); *O. latipes -G. aculeatus* (6078); *O. latipes -T. nigroviridis* (5531); *O. latipes -T. rubripes* (5946); *G. aculeatus -T. nigroviridis* (12088); *G. aculeatus -T. rubripes* (13415); *T. rubripes -T. nigroviridis* (13043).

Base Composition

Basic sequence information were retrieved by using Infoseq, an application of EMBOSS package (EMBOSS, Release 5.0; http://emboss.sourceforge.net/). Sequences with length less than hundred base pairs (bp) were excluded from further analysis.

CodonW (Version 1.4.4) was used to calculate the molar ratio of guanine plus cytosine (GC) of the entire genome as well as for the orthologous pairs and to calculate the di-nucleotides frequencies CpG, TpG and CpA.

Metabolic Rate and Genomic GC in Teleosts

Data on metabolism was principally retrieved from www.fishbase.org, as well as taxonomic classification, geographical distribution of teleostean fish. The mass specific metabolic rate (mg $kg^{-1}h^{-1}$) values were temperature-corrected utilizing the Boltzmann's factor [59]. The final dataset of 206 species were set into five habitats: polar, temperate, subtropical, tropical and deep-water.

The genomic GC levels of 149 teleostean fish were retrieved from current literature [60][40], representing 9 polar, 22 temperate, 48 subtropical and 70 tropical teleostean fish. Data on the genomic GC levels for fish living in the deep-water habitat were not available.

Statistical analyses were performed using the software StatView (Version 5.0).

Conclusion

After more than thirty years, the problem of the variation of base composition among genomes is an open question and still matter of debates between scientists supporting neutral hypotheses, based on stochastic events taking place during intracellular processes (i.e. DNA duplication, repair and recombination), or adaptive ones, based on the interaction of the organism with the surrounding environment. Hence, several forces have been proposed to drive the changes of base composition,i.e. GC content, affecting both coding and non coding regions: mutational bias [3-5], the breakpoints distribution [76], the bias gene conversion

(BGC) [12-14], the thermal stability [11] and the metabolic rate [30,31]. The aim of the present work was to delineate which of these forces may plays a more relevant evolutionary role taking into account that, most probably, several factors interact and converge on shaping the genome base composition, and which of them could be extended to the all living organisms.

At present, the mutational bias is contradicted by a huge amount of data showing that the AT pressure, which affects the genomes of both prokaryotes and eukaryotes, could not be able to produce the large range of GC variability observed among organisms [2,7,9,22,23].

Regarding the distribution of DNA breakpoints [76], it is well known that they are not uniformly distributed along the genome, but they occur more frequently in sections containing blocks of repetitive DNA [77] and are found in the proximal parts of chromosomes significantly more often than in the distal parts [78]. The latter point is in contradiction with the observation that telomeric regions are GC-rich regions showing a high gene concentration [79]. Moreover, a very small number of fragile sites are inherited in human, since the majority of them (94%) have been found in somatic cells [80]. The above data, and the fact that the isochore family patterns are conserved among mammalian genomes [18], raise uncertainties about the role of DNA breakpoints distribution in shaping the genome base composition. The same criticisms concern the BGC hypothesis. Indeed, also for the hotspot sites of recombination, the keystone of the BGC hypothesis, no phylogenetic conservation has been reported among mammals, even in closely related species such as chimpanzee and human [19], probably due to a loss of the hotspot recombination motifs in the human linage [81]. Interestingly, the BGC hypothesis found no support from the analysis of substitution patterns in bacteria [23].

Among the environmental factors proposed as driving force to shape GC content, temperature seemed a tempting candidate, owing to the thermal stability effect of the GC pairs on DNA, RNA and proteins [11]. It is worth to recall here that vertebrates show two different kind of genome evolution: the transitional and the shifting mode [17,82-85]. The former was observed comparing “cold-and warm-blooded vertebrates” (a terminology used to stress the fact that, considering temperature, both environmental and body temperature should be taken in to account [21], mainly amphibians versus mammals, observing an increment of the genome heterogeneity with temperature due to a non-uniform increment of the GC content affecting both coding and non-coding regions [11,86]. The latter, found among fish, is a complete compositional shift, like the one observed in bacteria. Indeed, mapping the “isochore families” [i.e. the DNA regions characterized by fairly homogeneous base composition [21]] along the chromosomes, no overlap at all was observed in compositionally far genomes, such as those of zebrafish and fugu, showing an average GC content of 36.9% and 44.0%, respectively [87]. However, considering the habitats of zebrafish and fugu [tropical (18°-24°C) and temperate (9°-13°C), respectively] a negative correlation was suggested, in contrast with the positive one expected in the light of the thermal stability hypothesis. In fact, analyzing genome composition of more than 200 teleostean fish [60], grouped according to their habitat, a negative trend between environmental temperature and GC content was found (Figure 2). This correlation is barely or not at all affected by the temperature dependence of the methylation/deamination process, as coming out by the analysis of CpG and derivative doublets in the intronic sequences of five completely sequenced genomes (Figure 7). Moreover, collecting the available data produced by several laboratories on the oxygen consumption on several teleostean species, also the MR showed a negative correlation (Fig.1), close to that observed in the case of the genomic GC content. Crossing data about MR and genomic GC content a positive correlation between the two variables was found (Figure 3), hence giving support to the metabolic rate hypothesis [30,31]. In this context it should be called to mind that in teleost fish a negative correlation has been reported between genome size and genomic GC content [60,88]. The finding fits the framework of the metabolic rate hypothesis, since it has been reported that the intron size is negatively correlated with the gene expression level [37,38,89].

It is worth to stress that grouping fish according to their habitats, instead of making a correlation with a single habitat-representative parameter like temperature, let us to detect outliers within each habitat. Indeed, habitats are, generally speaking, not delimited by boundaries, and in several cases it could be not easy to assign a given species to one habitat only. Moreover, regarding the metabolic rate, species can easily and quickly adapt to different aquatic parameters (like salinity, available O_2, pressure). On the contrary, the adaptation of the genomic GC content to the environment is expected to be a very slow process. Indeed, in the case of metabolic rate significant differences among habitats were found, whereas regarding the GC content two "main blocks" were found, the Polar/Temperate and the Sub-tropical/Tropical. Unfortunately, no data were available about the genomic GC content of deep-water fish. However, considering their MR (Figure 1), we predict that on the average the genomic GC content should be rather low. Data on the GC content of microbial communities living along the seawater column further supported the above prediction, stressing the link between available O_2 in the environment and genomic GC content [29].

In conclusion, an interesting inference of both the results obtained on the microbial communities living along the seawater column [29], and those on the fish genomes analysis according to habitats [55], is that the environmental physico-chemical parameters, independently from the phylogenetic relationship [55,56], affect the metabolic rate and could act as a selective factors shaping the genome base composition towards a converging "best-fit GC content". This scenario opens a possible answer to the challenging question "are there any neutral sites in the genomes of bacteria?" [2]. Indeed, if within each habitat the specific physico-chemical parameters define a threshold for the "best-fit GC content" of a given population, the changing of these parameters from habitat to habitat would affect the threshold of the negative, or purifying, selection mechanism. The shift of the of threshold for the "best-fit GC content" could hence account for the genome compositional shift observed comparing teleostean fish living in different habitats [55].

Acknowledgment

Thanks are due to Massimo Di Giulio, for critical discussion, and to Giorgio Bernardi and Annalisa Varriale, for kindly providing sketch of methylation in fish.

References

[1] Sueoka, N. (1961) Correlation between base composition of deoxyribonucleic acid and amino acid composition of proteins. *Proc. Natl. Acad. Sci. USA*, 47, 1141-1149.

[2] Rocha, E.P. and Feil, E.J. (2010) Mutational patterns cannot explain genome composition: are there any neutral sites in the genomes of bacteria? *PLoS Genet*, 6.

[3] Freese, E. (1962) On the evolution of base composition of DNA. *J. Theor. Biol.*, 3, 82:101.

[4] Sueoka, N. (1962) On the genetic basis of variation and heterogeneity of DNA base composition. *Proc. Natl. Acad. Sci. USA*, 48, 582-592.

[5] Lobry, J.R. and Sueoka, N. (2002) Asymmetric directional mutation pressures in bacteria. *Genome Biol*, 3, research0058-research0058.14.

[6] Sueoka, N. (1988) Directional mutation pressure and neutral molecular evolution. *Proc Natl Acad Sci U S A*, 85, 2653-2657.

[7] Sargentini, N.J. and Smith, K.C. (1994) DNA sequence analysis of spontaneous and gamma-radiation (anoxic)-induced lacId mutations in Escherichia coli umuC122::Tn5: differential requirement for umuC at G.C vs. A.T sites and for the production of transversions vs. transitions. *Mutat Res*, 311, 175-189.

[8] Lander, E.S., Linton, L.M., Birren, B., Nusbaum, C., Zody, M.C., Baldwin, J., Devon, K., Dewar, K., Doyle, M., FitzHugh, W. *et al.* (2001) Initial sequencing and analysis of the human genome. *Nature*, 409, 860-921.

[9] Alvarez-Valin, F., Lamolle, G. and Bernardi, G. (2002) Isochores, GC3 and mutation biases in the human genome. *Gene*, 300, 161-168.

[10] Lynch, M. (2007) *The Origins of Genome Architecture.* Sinauer Associated Inc., Sunderland, MA (USA).

[11] Bernardi, G. (2004) *Sructural and Evolutionary Genomics. Natural Selection in Genome Evolution.* Elsevier, Amsterdam.

[12] Eyre-Walker, A. (1993) Recombination and mammalian genome evolution. *Proc. R. Soc. Lond. B*, 252, 237-243.

[13] Galtier, N. and Duret, L. (2007) Adaptation or biased gene conversion? Extending the null hypothesis of molecular evolution. *Trends Genet*, 23, 273-277.

[14] Duret, L. and Galtier, N. (2009) Biased gene conversion and the evolution of mammalian genomic landscapes. *Annu Rev Genomics Hum Genet*, 10, 285-311.

[15] Birdsell, J.A. (2002) Integrating genomics, bioinformatics, and classical genetics to study the effects of recombination on genome evolution. *Mol Biol Evol*, 19, 1181-1197.

[16] Bernardi, G., Mouchiroud, D. and Gautier, C. (1988) Compositional patterns in vertebrate genomes: conservation and change in evolution. *J Mol Evol*, 28, 7-18.

[17] D'Onofrio, G., Jabbari, K., Musto, H., Alvarez-Valin, F., Cruveiller, S. and Bernardi, G. (1999) Evolutionary genomics of vertebrates and its implications. *Ann N Y Acad Sci*, 870, 81-94.

[18] Costantini, M., Cammarano, R. and Bernardi, G. (2009) The evolution of isochore patterns in vertebrate genomes. *BMC Genomics*, 10, 146.

[19] Ptak, S.E., Hinds, D.A., Koehler, K., Nickel, B., Patil, N., Ballinger, D.G., Przeworski, M., Frazer, K.A. and Paabo, S. (2005) Fine-scale recombination patterns differ between chimpanzees and humans. *Nat Genet*, 37, 429-434.

[20] Duret, L. and Galtier, N. (2009) Comment on "Human-specific gain of function in a developmental enhancer". *Science*, 323, 714; author reply 714.

[21] Bernardi, G., Olofson, B., Filipski, J., Zerial, M., Salinas, J., Cuny, G., Meunier-Rotival, M. and Rodier, F. (1985) The mosaic genome of warm-blooded vertebrates. *Science*, 228, 953-958.

[22] Hershberg, R. and Petrov, D.A. (2010) Evidence that mutation is universally biased towards AT in bacteria. *PLoS Genet*, 6.

[23] Hildebrand, F., Meyer, A. and Eyre-Walker, A. (2010) Evidence of selection upon genomic GC-content in bacteria. *PLoS Genet*, 6.

[24] Rocha, E.P. and Danchin, A. (2002) Base composition bias might result from competition for metabolic resources. *Trends Genet*, 18, 291-294.

[25] Woolfit, M. and Bromham, L. (2003) Increased rates of sequence evolution in endosymbiotic bacteria and fungi with small effective population sizes. *Mol Biol Evol*, 20, 1545-1555.

[26] Foerstner, K.U., von Mering, C., Hooper, S.D. and Bork, P. (2005) Environments shape the nucleotide composition of genomes. *EMBO Rep*.

[27] Musto, H., Naya, H., Zavala, A., Romero, H., Alvarez-Valin, F. and Bernardi, G. (2006) Genomic GC level,optimal growth temperature, and genome size in prokaryotes. *Biochem Biophys Res Commun*, 347, 1-3.

[28] Naya, H., Romero, H., Zavala, A., Alvarez, B. and Musto, H. (2002) Aerobiosis increases the genomic guanine plus cytosine content (GC%) in prokaryotes. *J Mol Evol*, 55, 260-264.

[29] Romero, H., Pereira, E., Naya, H. and Musto, H. (2009) Oxygen and guanine-cytosine profiles in marine environments. *J Mol Evol*, 69, 203-206.

[30] Vinogradov, A.E. (2001) Bendable genes of warm-blooded vertebrates. *Mol Biol Evol*, 18, 2195-2200.

[31] Vinogradov, A.E. (2005) Noncoding DNA, isochores and gene expression: nucleosome formation potential. *Nucleic Acids Res*, 33, 559-563.

[32] Vinogradov, A.E. and Anatskaya, O.V. (2006) Genome size and metabolic intensity in tetrapods: a tale of two lines. *Proc Biol Sci*, 273, 27-32.

[33] Privalov, P.L. and Gill, S.J. (1988) Stability of protein structure and hydrophobic interaction. *Adv Protein Chem*, 39, 191-234.

[34] D'Onofrio, G., Jabbari, K., Musto, H. and Bernardi, G. (1999) The correlation of protein hydropathy with the base composition of coding sequences [In Process Citation]. *Gene*, 238, 3-14.

[35] Eyre-Walker, A. and Hurst, L.D. (2001) The evolution of isochores. *Nat Rev Genet*, 2, 549-555.

[36] Jabbari, K., Cruveiller, S., Clay, O. and Bernardi, G. (2003) The correlation between GC3 and hydropathy in human genes. *Gene*, 317, 137-140.

[37] Versteeg, R., van Schaik, B.D., van Batenburg, M.F., Roos, M., Monajemi, R., Caron, H., Bussemaker, H.J. and van Kampen, A.H. (2003) The human transcriptome map reveals extremes in gene density, intron length, GC content, and repeat pattern for domains of highly and weakly expressed genes. *Genome Res*, 13, 1998-2004.

[38] Arhondakis, S., Auletta, F., Torelli, G. and D'Onofrio, G. (2004) Base composition and expression level of human genes. *Gene*, 325, 165-169.

[39] D'Onofrio, G., Ghosh, T.C. and Saccone, S. (2007) Different functional classes of genes are characterized by different compositional properties. *FEBS Lett*, 581, 5819-5824.

[40] Varriale, A. and Bernardi, G. (2006) DNA methylation and body temperature in fishes. *Gene*, 385, 111-121.

[41] Shen, J.C., Rideout, W.M., 3rd and Jones, P.A. (1994) The rate of hydrolytic deamination of 5-methylcytosine in double-stranded DNA. *Nucleic Acids Res*, 22, 972-976.

[42] Brown, J.H., Gillooly, J.F., Allen, A.P., Savage, V.M. and West, G.B. (2004) Toward a metabolic theory of ecology. *Evcology*, 85, 1771-1789.

[43] McGaughran, A. and Holland, B.R. (2010) Testing the effect of metabolic rate on DNA variability at the intra-specific level. *PLoS One*, 5, e9686.

[44] Kleiber, M. (1932) Body size and metabolism. *Hilgardia*, 6, 315-353.

[45] West, G.B., Brown, J.H. and Enquist, B.J. (1997) A general model for the origin of allometric scaling laws in biology. *Science*, 276, 122-126.

[46] West, G.B., Brown, J.H. and Enquist, B.J. (1999) The fourth dimension of life: fractal geometry and allometric scaling of organisms. *Science*, 284, 1677-1679.

[47] Bokma, F. (2004) Evidence against universal metabolic allometry. *Funct. Ecol.*, 18.

[48] Glazier, D.S. (2005) Beyond the '3/4-power law': variation in the intra-and interspecific scaling of metabolic rate in animals. *Biol Rev Camb Philos Soc*, 80, 611-662.

[49] White, C.R., Phillips, N.F. and Seymour, R.S. (2006) The scaling and temperature dependence of vertebrate metabolism. *Biol Lett*, 2, 125-127.

[50] Seibel, B.A. and Drazen, J.C. (2007) The rate of metabolism in marine animals: environmental constraints, ecological demands and energetic opportunities. *Philos Trans R Soc Lond B Biol Sci*, 362, 2061-2078.

[51] Makarieva, A.M., Gorshkov, V.G., Li, B.L., Chown, S.L., Reich, P.B. and Gavrilov, V.M. (2008) Mean mass-specific metabolic rates are strikingly similar across life's major domains: Evidence for life's metabolic optimum. *Proc Natl Acad Sci U S A*, 105, 16994-16999.

[52] Killen, S.S., Atkinson, D. and Glazier, D.S. (2010) The intraspecific scaling of metabolic rate with body mass in fishes depends on lifestyle and temperature. *Ecol Lett*, 13, 184-193.

[53] Kolokotrones, T., Van, S., Deeds, E.J. and Fontana, W. (2010) Curvature in metabolic scaling. *Nature*, 464, 753-756.

[54] Glazier, D.S. (2010) A unifying explanation for diverse metabolic scaling in animals and plants. *Biol Rev Camb Philos Soc*, 85, 111-138.

[55] Uliano, E., Chaurasia, A., Bernà, L., Agnisola, C. and D'Onofrio, G. (2010) Metabolic rate and genomic GC. What we can learn from teleost fish. *Marine Genomics*, 3, 29-34. doi:10.1016/j.margen.2010.02.001.

[56] Clarke, A. and Johnstone, N.M. (1999) Scaling of metabolic rate with body mass and temperature in teleost fish. *Journal of Animal Ecology*, 68, 893-905.

[57] Hodkinson, I.D. (2003) Metabolic cold adaptation in arthropods: a smaller-scale perspective. *Functional Ecology*, 17, 562-572.

[58] Karamushko, L.I. (2001) Metabolic Adaptation of Fish at High Latitudes. *Doklady Biological Sciences*, 379, 359-361.

[59] Gillooly, J.F., Brown, J.H., West, G.B., Savage, V.M. and Charnov, E.L. (2001) Effects of size and temperature on metabolic rate. *Science*, 293, 2248-2251.

[60] Bucciarelli, G., Bernardi, G. and Bernardi, G. (2002) An ultracentrifugation analysis of two hundred fish genomes. *Gene*, 295, 153-162.

[61] Schildkraut, C.L., Marmur, J. and Doty, P. (1962) Determination of the base composition of deoxyribonucleic acid from its buoyant density in CsCl. *J. Mol. Biol.*, 4, 430-443.

[62] Josse, J., Kaiser, A.D. and Kornberg, A. (1961) Enzymatic synthesis of deoxyribonucleic acid:VII. Frequencies of nearest neighbor base sequences in deoxyribonucleic acid. *J. Biol. Chem.*, 236, 864-875.

[63] Bird, A.P. (1980) DNA methylation and the frequency of CpG in animal DNA. *Nucleic Acids Res*, 8, 1499-1504.

[64] Lindahl, T., Karran, P. and Wood, R.D. (1997) DNA excision repair pathways. *Curr Opin Genet Dev*, 7, 158-169.

[65] Scarano, E., Iaccarino, M., Grippo, P. and Parisi, E. (1967) The heterogeneity of thymine methyl group origin in DNA pyrimidine isostichs of developing sea urchin embryos. *Proc Natl Acad Sci U S A*, 57, 1394-1400.

[66] Fryxell, K.J. and Zuckerkandl, E. (2000) Cytosine deamination plays a primary role in the evolution of mammalian isochores. *Mol Biol Evol*, 17, 1371-1383.

[67] Duret, L. and Galtier, N. (2000) The Covariation Between TpA Deficiency, CpG Deficiency, and G+C Content of Human Isochores Is Due to a Mathematical Artifact. *Mol Biol Evol*, 17, 1620-1625.

[68] Jabbari, K., Cacciò, S., Pais de Barros, J.P., Desgres, J. and Bernardi, G. (1997) Evolutionary changes in CpG and methylation levels in the genome of vertebrates. *Gene*, 205, 109-118.

[69] Jabbari, K. and Bernardi, G. (2004) Cytosine methylation and CpG, TpG (CpA) and TpA frequencies. *Gene*, 333, 143-149.

[70] Gardiner-Garden, M. and Frommer, M. (1987) CpG islands in vertebrate genomes. *J Mol Biol*, 196, 261-282.

[71] Larsen, F., Gundersen, G., Lopez, R. and Prydz, H. (1992) CpG islands as gene markers in the human genome. *Genomics*, 13, 1095-1107.

[72] Purugganan, M. and Wessler, S. (1992) The splicing of transposable elements and its role in intron evolution. *Genetica*, 86, 295-303.

[73] Dasilva, C., Hadji, H., Ozouf-Costaz, C., Nicaud, S., Jaillon, O., Weissenbach, J. and Roest Crollius, H. (2002) Remarkable compartmentalization of transposable elements and pseudogenes in the heterochromatin of the *Tetraodon nigroviridis* genome. *Proc Natl Acad Sci U S A*, 99, 13636-13641.

[74] Guryev, V., Koudijs, M.J., Berezikov, E., Johnson, S.L., Plasterk, R.H., van Eeden, F.J. and Cuppen, E. (2006) Genetic variation in the zebrafish. *Genome Res*, 16, 491-497.

[75] Lim, P.S., Hardy, K., Bunting, K.L., Ma, L., Peng, K., Chen, X. and Shannon, M.F. (2009) Defining the chromatin signature of inducible genes in T cells. *Genome Biol*, 10, R107.

[76] Lemaitre, C., Zaghloul, L., Sagot, M.F., Gautier, C., Arneodo, A., Tannier, E. and Audit, B. (2009) Analysis of fine-scale mammalian evolutionary breakpoints provides new insight into their relation to genome organisation. *BMC Genomics*, 10, 335.

[77] Bovero, S., Hankeln, T., Michailova, P., Schmidt, E. and Sella, G. (2002) Nonrandom chromosomal distribution of spontaneous breakpoints and satellite DNA clusters in two geographically distant populations of *Chironomus riparius* (Diptera: Chironomidae). *Genetica*, 115, 273-281.

[78] Michailova, P., Ilkova, J., Hankeln, T., Schmidt, E.R., Selvaggi, A., Zampicinini, G. and Sella, G. (2009) Somatic breakpoints, distribution of repetitive DNA and non-LTR retrotransposon insertion sites in the chromosomes of Chironomus piger Strenzke (Diptera, Chironomidae). *Genetica*, 135, 137-148.

[79] Saccone, S., De Sario, A., Della Valle, G. and Bernardi, G. (1992) The highest gene concentrations in the human genome are in telomeric bands of metaphase chromosomes. *Proc Natl Acad Sci U S A*, 89, 4913-4917.

[80] Abeysinghe, S.S., Stenson, P.D., Krawczak, M. and Cooper, D.N. (2004) Gross Rearrangement Breakpoint Database (GRaBD). *Hum Mutat*, 23, 219-221.

[81] Myers, S., Bowden, R., Tumian, A., Bontrop, R.E., Freeman, C., MacFie, T.S., McVean, G. and Donnelly, P. (2010) Drive against hotspot motifs in primates implicates the PRDM9 gene in meiotic recombination. *Science*, 327, 876-879.

[82] Bernardi, G. and Bernardi, G. (1990) Compositional transitions in the nuclear genomes of cold-blooded vertebrates. *J. Mol. Evol.*, 31, 282-293.

[83] Bernardi, G., Hughes, S. and Mouchiroud, D. (1997) The major compositional transitions in the vertebrate genome. *J Mol Evol*, 44, S44-51.

[84] Cruveiller, S., D'Onofrio, G. and Bernardi, G. (2000) The compositional transition between the genomes of cold-and warm-blooded vertebrates: codon frequencies in orthologous genes. *Gene*, 261, 71-83.

[85] D'Onofrio, G. and Ghosh, T.C. (2005) The compositional transition of vertebrate genomes: an analysis of the secondary structure of the proteins encoded by human genes. *Gene*, 345, 27-33.

[86] Clay, O., Arhondakis, S., D'Onofrio, G. and Bernardi, G. (2003) LDH-A and alpha-actin as tools to assess the effects of temperature on the vertebrate genome: some problems. *Gene*, 317, 157-160.

[87] Costantini, M., Auletta, F. and Bernardi, G. (2007) Isochore patterns and gene distributions in fish genomes. *Genomics*, 90, 364-371.

[88] Vinogradov, A.E. (1998) Genome size and GC-percent in vertebrates as determined by flow cytometry: the triangular relationship. *Cytometry*, 31, 100-109.

[89] Castillo-Davis, C.I., Mekhedov, S.L., Hartl, D.L., Koonin, E.V. and Kondrashov, F.A. (2002) Selection for short introns in highly expressed genes. *Nat Genet*, 31, 415-418.

[90] Altschul, S.F., Madden, T.L., Schaffer, A.A., Zhang, J., Zhang, Z., Miller, W. and Lipman, D.J. (1997) Gapped BLAST and PSI-BLAST: a new generation of protein database search programs. *Nucleic Acids Res*, 25, 3389-3402.

In: Fish Ecology
Editor: Sean P. Dempsey

ISBN 978-1-61324-282-7

Chapter 4

GENETIC STRUCTURE OF MASU SALMON, *ONCORHYNCHUS MASOU*

***Shigeru Kitanishi*[1] *and Toshiaki Yamamoto*[2]**
[1]College of Life Sciences, Ritsumeikan University, Nojihigashi 1-1-1, Kusatsu, Shiga 525-8577, Japan
[2]Department of Veterinary Nursing and Technology, Nippon Veterinary and Life Science University, Musashino, Tokyo 180-8602, Japan

ABSTRACT

Salmonid species have the potential to make a population genetic structure due to their homing behavior. However, the extent of genetic structuring would vary considerably because precision of homing varies within and among species. In addition to homing, several biotic and/or abiotic factors would also influence genetic structuring. A description of the genetic structure at various spatial scales and an understanding of the extent of genetic structuring could facilitate the identification of factors that affect genetic structuring. The knowledge of such factors is a fundamental requirement for the accurate inference of population dynamics, evolutionary processes, and conservation decisions. We focus on the factors that influence population genetic structuring and briefly describe the genetic structure of masu salmon (*Oncorhynchus masou*) populations at both regional and microgeographic scales. By analyzing mitochondrial DNA and microsatellite DNA variations, we found that masu salmon exhibit hierarchical genetic structuring and genetic differentiations not only at the regional scale but also at the microgeographic scale. These observations indicate that masu salmon would have the potential to make a population genetic structure at the microgeographic scale due to precise homing. Furthermore, it is also suggested that genetic structuring would be affected by several factors, such as refugia during glacial periods, ocean current, and dispersal patterns of each individual, and the effects of such factors may vary depending on the intended geographic scale. However, such intrinsic genetic structuring faces the danger of being eroded by anthropogenic effects, including habitat degradation, habitat fragmentation, and artificial release of hatchery-reared fish. In fact, the negative impact of damming would hold true for masu salmon. Our results indicate the possibility that the indigenous genetic structure that has been created over many years would be lost by human activities during short periods.

INTRODUCTION

Genetic similarities and differentiations among populations are the key sources of information to assess the evolutionary history of a species or population as well as the contemporary state and the future evolutionary potential (e.g., Frankham *et al.* 2004). Therefore, the knowledge of such factors is a fundamental requirement for the accurate inference of population dynamics, evolutionary processes, and conservation decisions. In addition, a description of the genetic structure at various spatial scales and an understanding of the extent of genetic structuring could facilitate the detection of evolutionary or ecological factors that influence genetic structuring and the quantification of their spatial scale of influence. In such a context, the genetic analyses of various species have revealed that the pattern and process of genetic structuring are strongly influenced by both biotic(e.g., dispersal ability, sex and life history, Blundell *et al.* 2002; Zardoya *et al.* 2004; Steele *et al.* 2009) and abiotic factors (i.e., environmental conditions, Krebs 2001; Hendry and Stearns 2004). The relative importance of these factor(s) may change in space and time, and some events could leave imprints for a long period (Hewitt 2000).

Salmonid fish are renowned for their extensive migration and homing behavior, in which an individual returns to their natal stream (Quinn and Dittman 1990; Hendry and Stearns 2004). Homing can facilitate the establishment of a distinct population and such isolation promotes genetic divergence and hence genetic structuring. However, precision of homing would vary considerably within and among species due to various selective pressures acting at different points in their journey (Hendry *et al.* 2004). Therefore, homing is viewed as a hierarchical phenomenon. In addition, salmonid fish exhibit life-history polymorphism; an individual would diverge to the seaward-migratory (anadromous) form or the freshwater resident form. The anadromous form has a larger body size than the resident form and undergoes long-distance migration, whereas the resident form remains in freshwater throughout its entire life. Because of the difference in swimming ability between the two forms (Peake *et al.* 1997; Kinnison *et al.* 2003), such life-history polymorphism can also affect genetic structuring (DeWoody and Avise 2000; Tonteri *et al.* 2007). Furthermore, several biotic and abiotic factors can influence genetic structuring. For example, ocean current, water temperature, and refugia during the ice age must have contributed to the limited distribution and migration patterns (e.g., Koljonen *et al.* 1999; Bernatchez 2001; Brykov *et al.* 2003; Polyakova *et al.* 2006). It is clear that several intrinsic and extrinsic factors would affect simultaneously the genetic structuring of salmonid fish and as a consequence, the genetic structure would vary considerably among species and populations (Hendry *et al.* 2004). A description of the genetic structure at various spatial scales and an understanding of the extent of genetic structuring could facilitate the identification of the factors that affect genetic structuring. Moreover, knowledge of the contribution of such factors is a fundamental requirement for the accurate inference of population dynamics, evolutionary processes, and conservation decisions.

In this chapter, we focus on the factors that influence the genetic structure of masu salmon (*Oncorhynchus masou*) populations. First, we briefly describe the life-history characteristics and genetic structures of masu salmon populations in Hokkaido, the northernmost island of Japan. Then, we discuss the factor(s) that influence genetic structuring at regional and microgeographic scales. Our aim is to assess the factors that could affect the genetic structuring of masu salmon populations at both spatial scales. We also would like to determine the implications of the conservation of the genetic structure of masu salmon populations.

CHARACTERISTICS OF MASU SALMON

Masu salmon is one of the species of the Pacific salmon, genus*Oncorhynchus*, and considered to be an ancestral species in this genus according to phylogenetic analysis (Mckay *et al.* 1996; Kitano *et al.* 1997; Oohara *et al.* 1997; Osinov and Lebedev 2000). Masu salmon is distributed only in the Far East, including Russia, Japan, and Korea, surrounded by the Sea of Japan, the Pacific Ocean, and the Sea of Okhotsk (Kato 1991). Like other salmonid species, masu salmon exhibits life-history polymorphism, with the anadromous form and the resident form coexisting with each other (Sugiwaka and Kojima 1984; Kato 1991). The number of anadromous individuals increases with increasing latitude, with most individuals becoming anadromous in Russia. In contrast, in the southern limit of their distribution, such as the Kyushu Island in Japan, only the resident form exists (Kato 1991). The two forms coexist in the central area, including Hokkaido and the northern part of Honshu Island. In those areas, life-history polymorphism is found only in males, that is, some males migrate to the sea while others remain in their natal river, whereas all females become migratory (Kato 1991; Kiso 1995). After seaward migration, they return to their natal stream in spring, with rare occasions of straying into non-natal rivers. They spawn in autumn and the spawning grounds are generally located in the upstream regions of rivers and tributaries. Masu salmon exhibits a promiscuous mating system where a single female simultaneously mates with multiple males, including both forms of males. All of the anadromous individuals are semelparous, whereas some resident males are iteroparous (Kato 1991; Kiso 1995).

Because of the importance of fisheries species, masu salmon have been well studied, including their propagation, habitat niche, breeding behavior, and population structure(e.g., Sugiwaka and Kojima 1984; Mayama 1990; Kiso 1995; Hayano *et al.* 2003; Yamamoto and Edo 2006; Hasegawa *et al.* 2010), in northern Japan. A few genetic studies were also conducted by using allozyme and mitochondrial DNA (mtDNA) analysis (Okazaki 1986; Suzuki *et al.* 2000; Edpalina *et al.* 2004). Significant genetic differentiations were observed between regions within northern Japan. However, probably because those studies included artificial effects (Allendorf and Phelps 1980; Okazaki 1986; Hansen and Mensberg 1998), several indistinct genetic clusterings were also observed in those studies. In addition, those studies focused on only the genetic structure at a broad spatial scale and therefore, knowledge of the genetic structure at the microgeographic scale, such as the within-river scale, and the factors that could have contributed to the genetic structure at those scales, has been limited. Here, to resolve the above issues, we analyzed the genetic structure of native masu salmon populations at two geographic scales: the whole area of Hokkaido as the regional scale and within the Atsuta River as the microgeographic scale. We used a native population in which there are no records of artificial releases of hatchery-reared fish to avoid anthropogenic effect(s) on the genetic structure, as mentioned above.

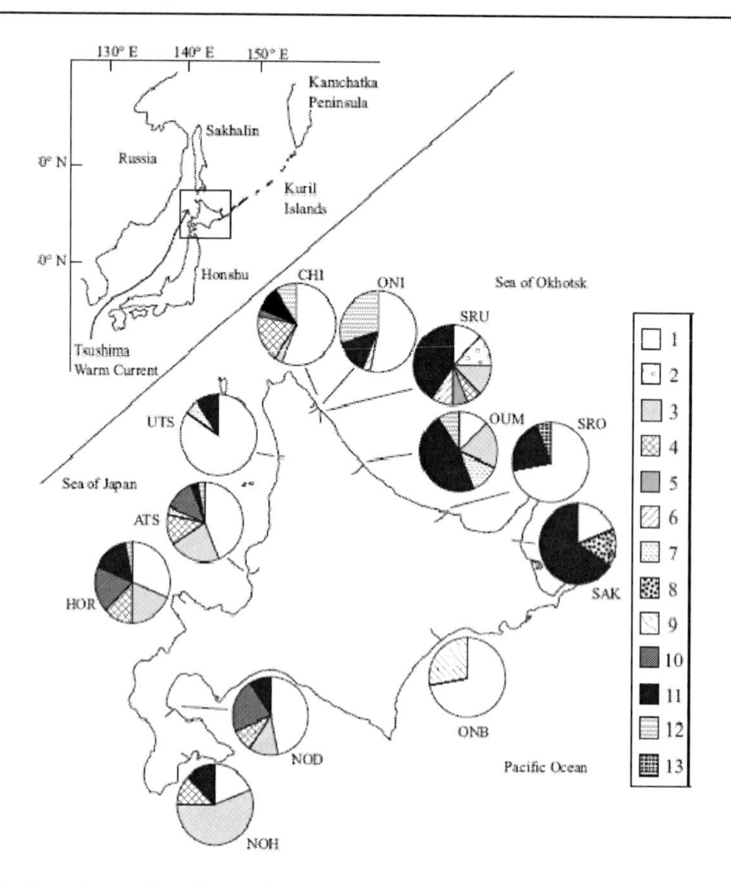

Figure 1. Distribution of sampling sites and of mtDNA haplotypes within Hokkaido.

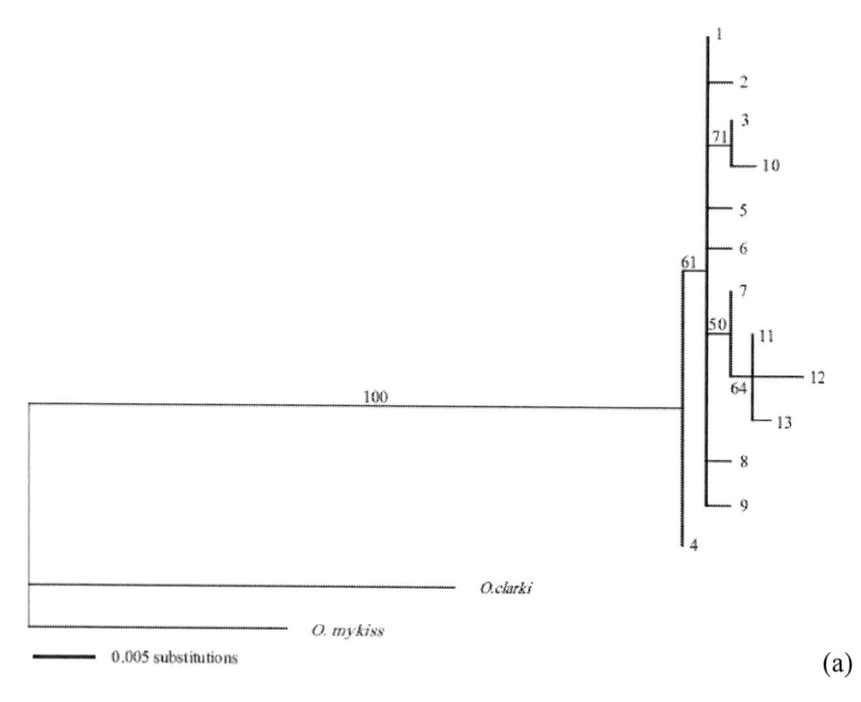

Figure 2. (continued).

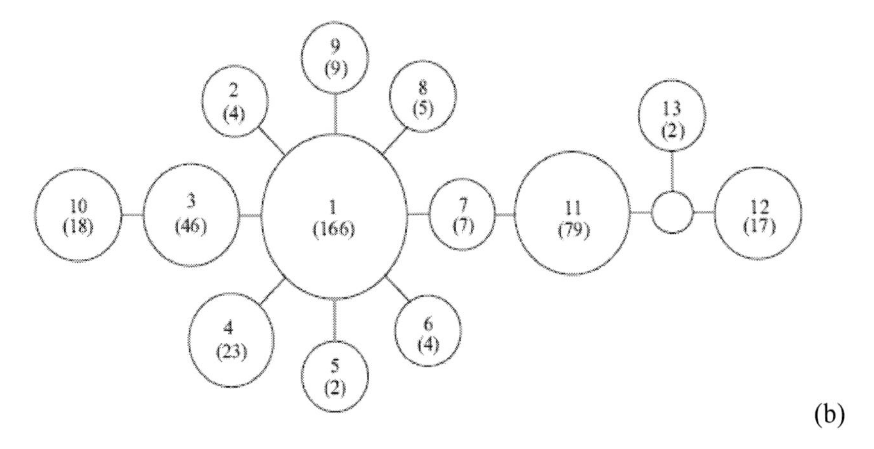

Figure 2. ML tree with 100 bootstraps (a) and network (b) of haplotypes of masu salmon inferredfrom 561 bp of the mtDNA ND5. In (b), circle size indicates the observed numbers of individualsgiven in parentheses.

Genetic Structure at Both Regional and Microgeographic Scales

First, to infer the genetic structure at the regional scale, we investigated 382 masu salmon individuals collected from 12 rivers around Hokkaido (Figure 1). We evaluated genetic differentiation and genetic structuring by using sequence analysis of the 5' half of the NADH dehydrogenase subunit 5 (ND5) gene in mtDNA(Kitanishi *et al.* 2007).

As a result, we identified a total of 13 variable nucleotide sites and 13 haplotypes in 561 bp of ND5 gene. The maximum-likelihood tree and the parsimony network (Clement *et al.* 2000) were used to infer the phylogenetic relationships among those haplotypes. Haplotype 4 may be an ancestral haplotype and frequently occurring haplotypes, such as haplotypes 1, 3, and 11, formed a stem of the network (Figure 2). Then, as a result of evaluation of the geographic distribution of these haplotypes, the frequently occurring haplotypes were found to have a wide distribution in Hokkaido, while the distribution of the other haplotypes was restricted regionally (Figure 1). In addition, the ratio of the haplotypes within each population showed regional bias. For example, haplotypes 11 and 12 commonly occurred in populations existing in the Sea of Okhotsk, while haplotypes 3 and 4 were found in the southern region of Hokkaido.

To infer whether masu salmon populations are structured or not, we first divided the populations into three groups: Sea of Japan group, Sea of Okhotsk group, and Pacific Ocean group, and then conducted a hierarchical analysis of molecular variance (AMOVA; Excoffier *et al.* 1992). We found significant genetic differentiations among the above three groups and the genetic variance at the between-groups level accounted for 10.7% of the total genetic variance. The haplotypes observed within each group also differed among the groups (all comparisons; $\chi^2 > 17.7$, $P < 0.001$). In addition, genetic differentiations at the between-populations (10.9%) and between-individuals (78.5%) levels were also significant, suggesting hierarchical genetic structuring. In fact, significant genetic differentiations, as estimated by using F_{ST}, were observed not only in populations of different groups (range: -0.024 to 0.537) but also in those of the same group (range: 0.013 to 0.341). A significant correlation between genetic and geographic distance (IBD) was observed as well, suggesting that gene flow between distant populations may occur less frequently. Moreover, to infer the population relationship, populations were clustered by the neighbor-joining (NJ) method (Saitou and Nei 1987) using D_A as a distance measure. As a result, populations located in southeastern Hokkaido were grouped into a single cluster, whereas other populations did not form any clear clusters (Figure 3). To infer whether a historical demographic event occurred or not, Tajima's D statistic (Tajima 1989), Fu's Fs (Fu 1997), and the mismatch distribution test (Rogers and Harpending 1992) were used. Those analyses suggested that a sudden expansion of population size occurred in the entire population and one population located in northernmost Hokkaido (CHI, Figure 1), whereas those in the other regions, including southeastern Hokkaido, did not experience such a phenomenon. The population expansion occurred 79,000 to 350,000 years ago in the entire population of Hokkaido and 15,000 to 68,000 years ago in CHI.

Then, to assess the genetic structure at within-river scale, we sampled 280 anadromous individuals from seven tributaries of the Atsuta River (Figure 4) and used eight microsatellite DNA loci(Kitanishi *et al.* 2009). This river is approximately 34 km long and located in mid-western Hokkaido. In this river, fishing was prohibited for approximately 40 years due to fisheries regulations and there are no artificial constructions, such as weirs and dams. Therefore, the fish samples of this river are less influenced by human activity and appropriate for the study of native genetic structure at the microgeographic scale.

As a result of genetic analyses based on microsatellite DNA polymorphism, significant genetic differentiation was observed among the seven tributaries within the Atsuta River (Fisher's exact test: $P< 0.001$; global $F_{ST} = 0.019$, $P< 0.001$). In addition, despite the very close geographic proximity, private alleles were observed within each population (range: 0 to 4). For pairwise comparisons, the genetic composition within each tributary sample differed between most of the population pairs, with 15 of 20 comparisons showing a significant difference after the false discovery rate (FDR) correction (Benjamini and Yekutieli 2001). Significant genetic differentiations estimated from F_{ST} values were also observed in nine cases. The IBD signal was also observed, suggesting that gene flow would be more frequent between neighboring tributaries (Figure 5). Principal component analysis illustrated a well-supported geographical basis of population relationships (Figure 6). These results indicate that masu salmon have the potential to make a genetic structure at the microgeographic scale (ca. < 21 km) and hierarchical genetic structuring occurs even at the within-river scale.

FACTORS INFLUENCING GENETIC STRUCTURING

The salient findings of our studies were as follows: (1) significant genetic differentiation and relatively clear genetic structuring were noted at both regional and microgeographic scales, (2) masu salmon have the potential to create indigenous populations even at the within-river scale (< 21 km), (3) hierarchical genetic structuring as inferred from genetic differentiation and IBD signals would occur from the regional-group level to the tributary level. Then why would masu salmon exhibit a clear genetic structure at various spatial scales? What factor(s) could lead to such genetic structuring for masu salmon populations?

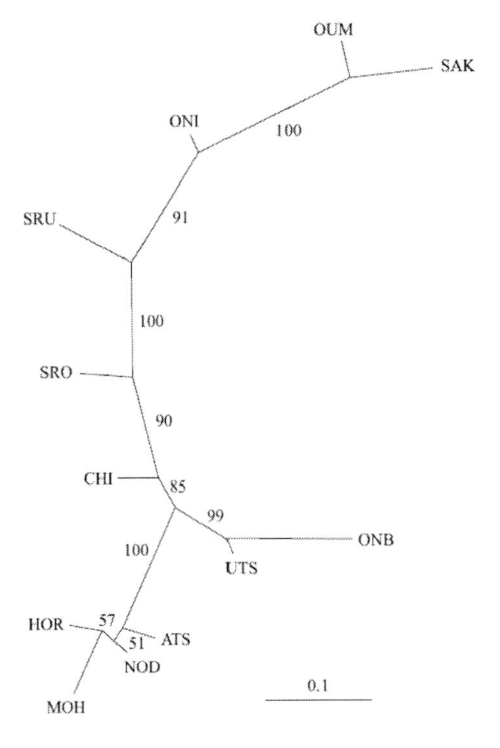

Figure 3. Unrooted NJ tree of 12 masu salmon populations, inferred from DAwith bootstrap values (n= 1000). Abbreviations, see Figure 1.

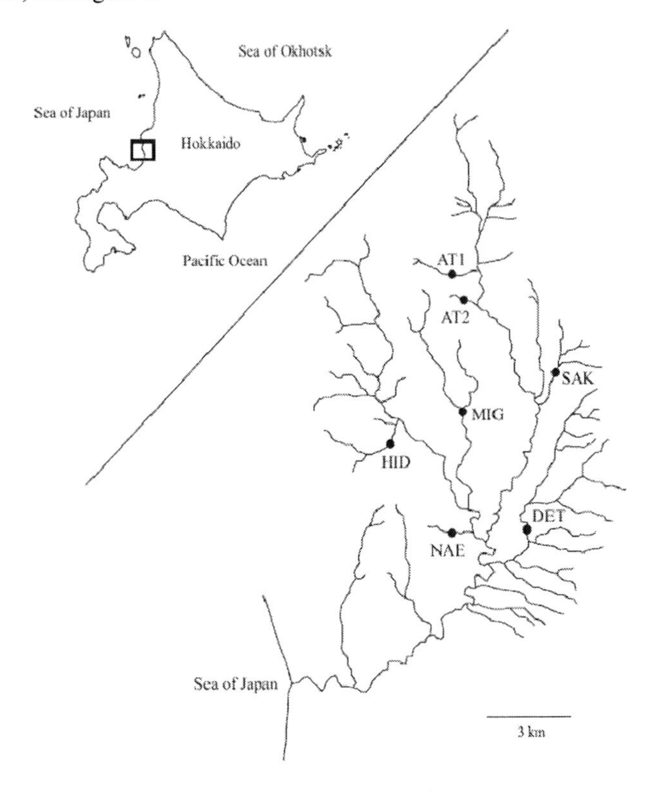

Figure 4. Distribution of seven sampling sites in th e Atsuta River.

Precise Homing of Masu Salmon

First of all, the precise homing behavior is an intrinsic feature of masu salmon. Genetic differentiations estimated from F_{ST} among river populations were relatively high. For instance, significant genetic differentiation and genetic clustering were not observed in chum salmon, *O. keta*, within a similar spatial scale (Okazaki 1990; Sato *et al.* 2004). Differences in environmental conditions between locations may increase with increasing geographic distance. Masu salmon populations in different locations show remarkable divergence in their life-history characteristics, such as body size, timing of migration, and breeding period (Kato 1991). Under such conditions, strays should have lower reproductive success than homers and therefore gene flow between distant populations may occur less frequently. In addition, our study also revealed that tributary populations also showed significant genetic differentiation even between proximal samples. Despite the fact that the degree of differentiation observed in our study (global $F_{ST} = 0.019$) is in the lower end of the ranges observed in earlier studies of within-river scale using microsatellite DNA($F_{ST} = 0.014$ - 0.082; Garant *et al.* 2000; Spidle *et al.* 2001; Primmer *et al.* 2006; Neville *et al.* 2007; Vähä *et al.* 2007), relatively clear genetic structuring and significant IBD signal was observed in the Atsuta River. Regardless of the superior swimming ability of the anadromous form (Peake *et al.* 1997; Kinnison *et al.* 2003), fine-scale genetic structuring was observed in the Atsuta River. Therefore, we confirm that the homing of masu salmon would be accurate even at the within-river level, and our results must represent the potential of masu salmon to make a genetic structure in the native condition.

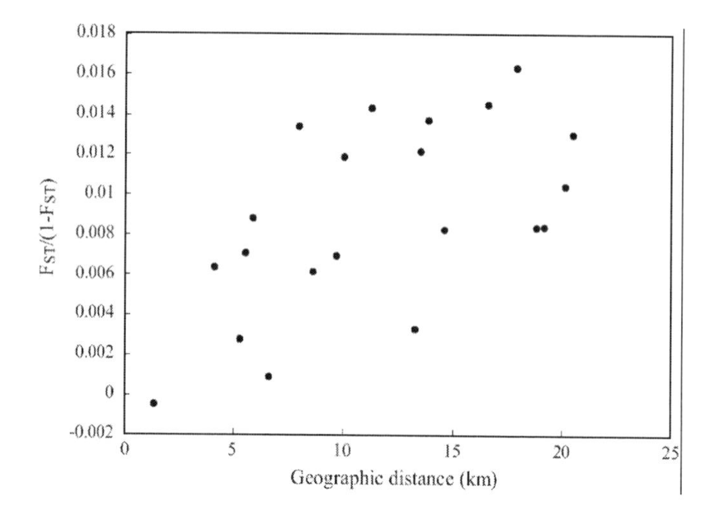

Figure 5. Correlation between genetic differentiation and geographic distance forseven samples in the Atsuta River (Mantel test: $p = 0.018$, $r = 0.622$).

So why does masu salmon have such precise homing and hence fine-scale genetic structure? We speculate that two nonexclusive life-history characteristics may contribute to this. The first and most important characteristic leading to fine-scale genetic structuring is the existence of the resident form within a population. Because of their low swimming ability (Taylor and Foote 1991; Peake *et al.* 1997; Kinnison *et al.* 2003), the cost of dispersal of the resident individuals may constrain their movements. Sakata et al. (2005) reported the high sedentary tendency of the fluvial form of masu salmon and the less than 30 m migration of more than half of the mobile individuals. The low genetic differentiations among anadromous populations compared to the resident populations have been reported in some fish species(DeWoody and Avise 2000; Tonteri *et al.* 2007). In the Atsuta River, approximately 30% of males become the resident form (Sugiwaka and Kojima 1984) and the resident form with high fidelity may facilitate genetic divergence even between neighboring tributaries. In addition, resident individuals cannot contribute to gene flow at the between-river level. Thus, the existence of the resident form per se may encourage genetic differentiation between different river populations. This is supported by the fact that salmonid fish with fine-scale genetic structure also have migratory dichotomy(e.g., Garant *et al.* 2000; Neville *et al.* 2007; Vähä *et al.* 2007) and wholly resident populations exhibit greater genetic differentiation than anadromous populations (Hendry *et al.* 2004).

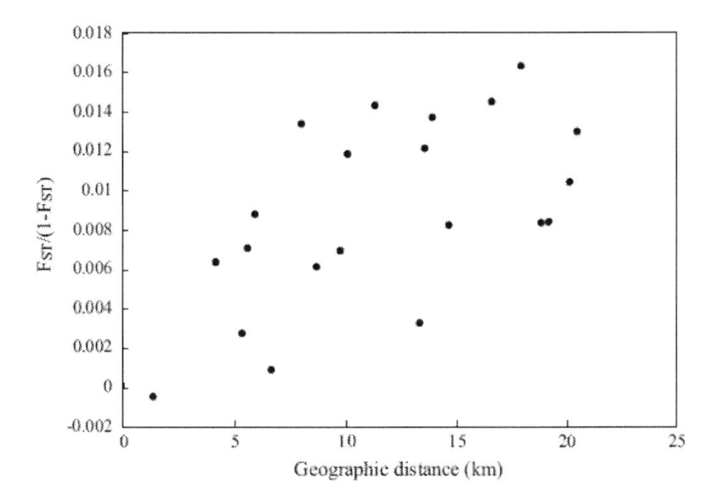

Figure 5. Correlation between genetic differentiation and geographic distance forseven samples in the Atsuta River (Mantel test: $p = 0.018$, $r = 0.622$).

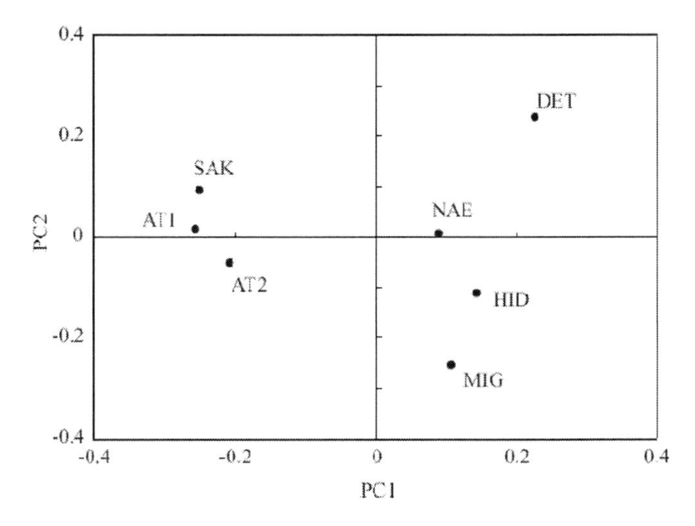

Figure 6. Results of principal components analysis of genetic variation amongsamples. PC1 explained 32% of the variation and PC2 explained 19%. Abbreviations, see Figure 4.

As the second characteristic, the duration of a juvenile's stay in its natal river may also contribute to precision of homing (Okazaki 1990; Utter 2001). It is thought that juvenile salmon learn the chemical characteristics of the water at their natal stream before seaward migration and later, the adults use these memories for homing (Quinn and Dittman 1990; Dittman and Quinn 1996). Juvenile masu salmon spend at least one year in their natal stream (Kato 1991) and it is thought that this period may be sufficient for imprinting. This assertion may well explain the fact that salmonid species without juvenile's stay, such as chum salmon and pink salmon, *O. gorbuscha*, exhibit higher straying rate than salmonids with juvenile's stay, such as sockeye salmon, *O. nerka*, and Atlantic salmon, *Salmo salar* L. (Quinn 1993; Hendry *et al.* 2004; Lin *et al.* 2008). However, salmonid fish with long-term stay also exhibit the two migratory forms. Further empirical research is needed to confirm this assertion.

Nevertheless, if the above two life-history characteristics are the key components for the determination of precision of homing, the spatial scale of the genetic structure observed in masu salmon may vary depending on the location, because these life histories vary dramatically among locations (Kato 1991). For example, there are no anadromous individuals in populations located at regions south of their distribution, whereas in the northern area, such as the Sakhalin, all females and a majority of males become anadromous (Kato 1991). This suggests that the spatial scale of the genetic structure may be larger in the northern area than in the southern area. Recently, Yu *et al.* (2010) offered an insight into this question. They investigated the genetic structure of a masu salmon population in the Far East, including northern Japan, Korea, and eastern Russia, by using ND5 gene in mtDNA and six microsatellite DNA loci, and found weaker genetic structuring in Russia than in Japan (Yu *et al.* 2010). Despite the sampling gaps and the inclusion of domesticated populations, their result would be concordant with our assumption of large-scale genetic structuring in the northern area.

Then, does such trend of spatial scale of genetic structuring hold true at the within-river scale? To answer this question, we primarily investigated the effects of sex and life history on the genetic structuring in the Atsuta River (Kitanishi *et al.* submitted) and revealed that both sex and migratory life history exhibit different patterns of dispersal at the microgeographic scale. Genetic differentiation in the Atsuta River can be mainly attributed to female philopatry, whereas anadromous males may be agents of gene flow between tributaries. In addition, resident individuals also exhibit significant genetic differentiations among tributary samples, indicating that resident individuals could, at least in part, contribute to genetic structuring. These results indicate that the variety of the spatial scale of genetic structuring, as mentioned above, may also be applicable to the within-river scale. This opens an interesting avenue for future research.

Refugia and Postglacial Colonization

At the regional scale, our study indicated that refugia during glaciations and the routes of postglacial colonization could contribute to the genetic structure of masu salmon populations in Hokkaido. It is well known that the distribution of refugia from ice cover and the postglacial dispersal may be of importance to the current genetic structure of salmonid fish that are distributed in Europe and North America(e.g., Koljonen *et al.* 1999; Mccusker *et al.* 2000; Bernatchez 2001; King *et al.* 2001).

Similar influences of glaciations may occur in salmonid fish inhabiting the Far East. In fact, it was reported that chum salmon inhabiting the Sea of Okhotsk experienced significant size reduction in the Pleistocene glaciations, whereas those inhabiting the southern part of the Sea did not (Brykov *et al.* 2003). Applying this to masu salmon, masu salmon may not breed in rivers located at the Sea of Okhotsk and the northern region of the Sea of Japan during glacial periods, because the water temperature preferred by masu salmon is higher than that preferred by chum salmon (Kato 1991; Salo 1991). The present distribution of masu salmon is up to 55° N, where the annual mean air temperature is ca. -6°C. In the glacial periods, this -6°C line is estimated to have been around northern Hokkaido (Matsusue *et al.* 2000). These suggest that the southeastern region of Hokkaido may serve as refugia for the masu salmon population in the ice age. After the ice age, colonization from this refugium area may occur in the northern region along the Tsushima Warm Current (Figure 1). This could also be supported by the results of genetic clustering (NJ tree) and demographic analyses, which indicated the occurrence of population size expansion in the Hokkaido populations. In addition, the fact that the southeastern Hokkaido populations share a presumably ancestral haplotype 4 may also support this scenario. Therefore, we conclude that the genetic structure of masu salmon at the regional spatial scale would strongly reflect the effects of refugium in the ice age, followed by the northward expansion in the postglacial periods. And the genetic composition within each population would have been changing graduallyowing to ongoing gene flow between neighboring populations.

Minimal Influence of Anthropogenic Effect(s)

The minimal influence of anthropogenic effect(s) could encourage the genetic structuring observed in our studies. The fact that the artificial release of hatchery-reared fish could lead to genetic introgression and hence indistinct genetic structure has been reported in various salmonids (e.g., Nielsen *et al.* 1999; Utter 2001; Caputo *et al.* 2004). In masu salmon, vast amounts of fish have been released in northern Japan since the 1880s (Mayama 1992) and similar consequences were also reported by previous genetic studies (Okazaki 1986; Suzuki *et al.* 2000). In addition, cultured fish would differ in their morphological, ecological, behavioral, and genetic features and poorly adapt to local conditions(Hindar *et al.* 1991; Jonsson and Jonsson 2006; Araki *et al.* 2008). If an interbreeding of wild salmon and hatchery ones were to take place, the reduced fitness of wild salmon would come from genetic introgression (Gross 1998; Araki *et al.* 2008). Even if the introgression did not occur, wild populations would be exposed to detrimental effects though direct and/or indirect competition for mates, food, and breeding site (Weber and Fausch 2003). However, little is known about the above issues in the masu salmon populations and thus, we must improve our understanding of the genetic interactions between wild and hatchery-produced salmon.

In addition, the absence of artificial construction, such as weirs and dams, can also contribute to genetic structuring, particularly at the within-river scale. It is well known that damming has several consequences, the most serious of which is habitat fragmentation. Damming prevents aquatic animals from migrating, particularly upward migration in river systems; thus, the upstream populations become isolated (Gehrke *et al.* 2002; Gosset *et al.* 2006). In addition, small populations resulting from the isolation would be exposed to a higher risk of extinction through genetic, environmental, and demographic stochasticity

(Frankham *et al.* 2004). The influence of such disruption would be particularly relevant to salmonid fish, because they require seaward and upward migration as part of their life cycle. In fact, such significant negative impacts as the reduction of population size, the loss of genetic diversity, and indistinct genetic structuring, have been reported in salmonid fish (e.g., Hansen and Mensberg 1998; Bouza *et al.* 1999; Morita and Yamamoto 2002; Yamamoto *et al.* 2004). In the Atsuta River, there is neither artificial construction nor release of domesticated fish. Therefore, such a condition could, in part, contribute to the clear genetic structuring observed in this study.

Our study provides some insights to understand the factors influencing genetic structuring at both regional and microgeographic scales. We revealed that masu salmon exhibit hierarchical genetic structuring and have the potential to make a clear genetic structure even at the fine scale due to precise homing. It is well known that masu salmon have highly variable morphology, behavior, and genetic composition at any geographical scale (e.g., Kato 1991; Kiso 1995; Kitanishi *et al.* 2007). These indigenous features may result from past adaptation to their habitat as well as reflect contemporary status and include future evolutionary potential (Frankham *et al.* 2004). However, these genetic structures would be in danger of being eroded by ongoing anthropogenic effects. Recently, to evaluate whether or not damming negatively affects masu salmon populations, we collected masu salmon samples from above and below an impassable dam in the Uryu River, northern Hokkaido, and analyzed microsatellite DNA polymorphisms (Kitanishi *et al.* unpublished data). Although genetic diversity indices, such as heterozygosity, did not differ between the populations of the two regions, loss of the observed number of alleles was revealed in populations of the above-dam region, indicating that loss of genetic diversity has been going on in populations of the above-dam region (Luikart and Cornuet 1998). In addition, comparisons of the genetic structures of the populations of this river and those of the Atsuta River demonstrated that loss of genetic diversity and depopulation as evaluated from the effective population size may occur not only in populations of the above-dam region but also in those of the below-dam region, probably because of habitat degradation in both regions associated with human activities. These findings suggest that the indigenous genetic structure that has been created over many years would be lost by human activities during short periods. From conservation perspective, it is clear that the conservation of indigenous features inherent to each location is important for the long-term preservation of masu salmon populations (e.g., Frankham *et al.* 2004). In addition, as the artificial release of hatchery-reared fish and habitat alterations are needed to sustain human lives, it is important to devise ways to reduce the negative influences of such human activities. To this end, an understanding of the genetic structure, including its factor(s), and the characterization of indigenous populations are important as they may provide valuable insights for conservation activities and management on a situational basis (Frankham *et al.* 2004; Kawamura *et al.* 2007).

CONCLUSION

Using genetic analyses, we have unraveled hierarchical genetic structuring and how biotic and abiotic factors interact and generate genetic structures of masu salmon populations at both regional and microgeographic scales. At the regional scale, refugia during the glacial periods, postglacial colonization, and ongoing gene flow were generally accounted for, whereas precise homing and dispersal patterns depending on intrinsic characteristics (i.e., sex and life histories) resulted in fine-scale genetic structuring. Despite the fact that genetic analysis was proven to be a robust method for the identification of the extent of genetic structuring, the evolutionary processes, and the conservation perspectives, such patterns of spatial genetic structure would interact with several other intrinsic and extrinsic attributes. Future studies aiming at more rigorous treatments of these attributes would lead to a more generalized understanding.

REFERENCES

Allendorf, F. W. and Phelps, S. R. (1980). Loss of genetic variation in a hatchery stock of Cutthroat trout. *Transactions of the American Fisheries Society,* 109, 537-543.

Araki, H., Berejikian, B. A., Ford, M. J. and Blouin, M. S. (2008). Fitness of hatchery-reared salmonids in the wild. *Evolutionary Applications,* 1, 342-355.

Benjamini, Y. and Yekutieli, D. (2001). The control of the false discovery rate in multiple testing under dependency. *Annals of Statistics,* 29, 1165-1188.

Bernatchez, L. (2001). The evolutionary history of brown trout (*Salmo trutta* L.) inferred from phylogeographic, nested clade, and mismatch analyses of mitochondrial DNA variation. *Evolution,* 55, 351-379.

Blundell, G. M., Ben-David, M., Groves, P., Bowyer, R. T. and Geffen, E. (2002). Characteristics of sex-biased dispersal and gene flow in coastal river otters: implications for natural recolonization of extirpated populations. *Molecular Ecology,* 11, 289-303.

Bouza, C., Arias, J., Castro, J., Sanchez, L. and Martinez, P. (1999). Genetic structure of brown trout, *Salmo trutta* L., at the southern limit of the distribution range of the anadromous form. *Molecular Ecology,* 8, 1991-2001.

Brykov, V. A., Polyakova, N. E. and Prokhorova, A. V. (2003). Phylogeographic analysis of chum salmon*Oncorhynchus keta* walbaum in Asian populations based on mtDNA variation. *Russian Journal of Genetics,* 39, 61-67.

Caputo, V., Giovannotti, M., Cerioni, P. N., Caniglia, M. L. and Splendiani, A. (2004). Genetic diversity of brown trout in central Italy. *Journal of Fish Biology,* 65, 403-418.

Clement, M., Posada, D. and Crandall, K. A. (2000). TCS: A computer program to estimate gene genealogies. *Molecular Ecology,* 9, 1657-1659.

DeWoody, J. A. and Avise, J. C. (2000). Microsatellite variation in marine, freshwater and anadromous fishes compared with other animals. *Journal of Fish Biology,* 56, 461-473.

Dittman, A. H. and Quinn, T. P. (1996). Homing in Pacific salmon: Mechanisms and ecological basis. *Journal of Experimental Biology*, 199, 83-91.

Edpalina, R. R., Yoon, M., Urawa, S., Kusuda, S., Urano, A. and Abe, S. (2004). Genetic variation in wild and hatchery populations of masu salmon (*Oncorhynchus masou*) inferred from mitochondrial DNA sequence analysis. *Fish Genetics and Breeding Science,* 34, 37-44.

Excoffier, L., Smouse, P. E. and Quattro, J. M. (1992). Analysis of molecular variance inferred from metric distances among DNA haplotypes: application to human mitochondrial DNA restriction data. *Genetics,* 131, 479-491.

Frankham, R., Ballow, J. D. and Briscoe, D. A. (2004). *A primer of conservation genetics* (Cambridge, Cambridge University Press.

Fu, Y.-X. (1997). Statistical tests of neutrality of mutations against population growth, hitchhiking and background selection. *Genetics,* 147, 915-925.

Garant, D., Dodson, J. J. and Bernatchez, L. (2000). Ecological determinants and temporal stability of the within-river population structure in Atlantic salmon. *Molecular Ecology,* 9, 615-628.

Gehrke, P. C., Gilligan, D. M. and Barwick, M. (2002). Changes in fish communities of the Shoalhaven River 20 years after construction of Tallowa Dam, Australia. *River Research and Applications,* 18, 265-286.

Gosset, C., Rives, J. and Labonne, J. (2006). Effect of habitat fragmentation on spawning migration of brown trout (*Salmo trutta* L.). *Ecology of Freshwater Fish,* 15, 247-254.

Gross, M. R. (1998). One species with two biologies: Atlantic salmon (*Salmo salar*) in the wild and in aquaculture. *Canadian Journal of Fisheries and Aquatic Sciences,* 55 (Suppl. 1), 131-144.

Hansen, M. M. and Mensberg, K.-L. D. (1998). Genetic differentiation and relationship between genetic and geographical distance in Danish sea trout (*Salmo trutta* L.) populations. *Heredity,* 81, 493-504.

Hasegawa, K., Yamamoto, T. and Kitanishi, S. (2010). Habitat niche separation of the nonnative rainbow trout and native masu salmon in the Atsuta River, Hokkaido, Japan. *Fisheries Science,* 76, 251-256.

Hayano, H., Miyakoshi, Y., Nagata, M., Sugiwaka, K. and Irvine, J. R. (2003). Age composition of masu salmon smolts in northern Japan. *Journal of Fish Biology,* 62, 237-241.

Hendry, A. P., Castric, V., Kinnison, M. T. and Quinn, T. P. (2004). The evolution of philopatry and dispersal. In: A. P. Hendry and S. C. Stearns (eds.), *Evolution Illuminated* (pp. 52-91). New York, Oxford university press.

Hendry, A. P. and Stearns, S. C. (2004). *Evolution Illuminated.* New York, Oxford University Press.

Hewitt, G. (2000). The genetic legacy of the Quaternary ice ages. *Nature,* 405, 907-913.

Hindar, K., Ryman, N. and Utter, F. (1991). Genetic-Effects of Cultured Fish on Natural Fish Populations. *Canadian Journal of Fisheries and Aquatic Sciences,* 48, 945-957.

Jonsson, B. and Jonsson, N. (2006). Cultured Atlantic salmon in nature: a review of their ecology and interaction with wild fish. *Ices Journal of Marine Science,* 63, 1162-1181.

Kato, F. (1991). Life histories of masu and amago salmon (*Oncorhynchus masou* and *Oncorhynchus rhodurus*). In: C. Groot and L. Margolis (eds.), *Pacific Salmon life histories* (pp. 447-520). Vancouver, UBC Press.

Kawamura, K., Kubota, M., Furukawa, M. and Harada, Y. (2007). The genetic structure of endangered indigenous populations of the amago salmon, *Oncorhynchus masou ishikawae*, in Japan. *Conservation Genetics,* 8, 1163-1176.

King, T. L., Kalinowski, S. T., Schill, W. B., Spidle, A. P. and Lubinski, B. A. (2001). Population structure of atlantic salmon: a range-wide perspective from microsatellite DNA variation. *Molecular Ecology,* 10, 807-821.

Kinnison, M. T., Unwin, M. J. and Quinn, T. P. (2003). Migratory costs and contemporary evolution of reproductive allocation in male chinook salmon. *Journal of Evolutionary Biology,* 16, 1257-1269.

Kiso, K. (1995). The life history of masu salmon*Oncorhynchus masou* originated form rivers of the pacific coast of northern Honshu, Japan. *Bulletin of the National Research Institute of Fisheries Science,* 7, 1-188.

Kitanishi, S., Edo, K., Yamamoto, T., Azuma, N., Hasegawa, O. and Higashi, S. (2007). Genetic structure of masu salmon (*Oncorhynchus masou)* populations in Hokkaido, northernmost Japan, inferred from mitochondrial DNA variation. *Journal of Fish Biology,* 71, 437-452.

Kitanishi, S., Yamamoto, T. and Higashi, S. (2009). Microsatellite variation reveals fine-scale genetic structure of masu salmon, *Oncorhynchus masou,* within the Atsuta River. *Ecology of Freshwater Fish,* 18, 65-71.

Kitano, T., Matsuoka, N. and Saito, N. (1997). Phylogenetic relationship of the genus*Oncorhynchus* species inferred from nuclear and mitochondrial markers. *Genes and Genetic Systems,* 72, 25-34.

Koljonen, M.-L., Jansson, H., Paaver, T., Vasin, O. and Koskiniemi, J. (1999). Phylogeographic lineages and differentiation pattern of atlantic salmon in the Baltic Sea with management implications. *Canadian Journal of Fisheries and Aquatic Science,* 56, 1766-1780.

Krebs, C. J. (2001). *Ecology: the experimental analysis of distribution and abundance* (5th Edition). San Francisco, Addison Wesley.

Lin, J., Quinn, T. P., Hilborn, R. and Hauser, L. (2008). Fine-scale differentiation between sockeye salmon ecotypes and the effect of phenotype on straying. *Heredity,* 101, 341-350.

Luikart, G. and Cornuet, J. M. (1998). Empirical evaluation of a test for identifying recently bottlenecked populations from allele frequency data. *Conservation Biology,* 12, 228-237.

Matsusue, K., Fujiwara, O. and Sueyoshi, T. (2000). Climatic condition in the last glacial maximum in Japan. *Saikuru Kikou Gihou,* 6, 93-103.

Mayama, H. (1990). Masu salmon propagation in hokkaido, japan. Bulletin of the Institute of Zoology, *Academia Sinica,* 29, 95-104.

Mayama, H. (1992). Studies on the freshwater life and propagation technology of masu salmon, *Oncorhuynchus masou* (Brevoort). *Scientific Reports of the Hokkaido Salmon Hatchery,* 46, 1-156.

Mccusker, M. R., Parkinson, E. and Taylor, E. B. (2000). Mitochondrial DNA variation in rainbow trout (*Oncorhynchus mykiss*) across its native range: testing biogeographical hypotheses and their relevance to consevation. *Molecular Ecology,* 9, 2089-2108.

Mckay, S. J., Devlin, R. H. and Smith, M. J. (1996). Phylogyny of pacific salmon and trout based on growth hormon type-2 and mitochondrial NADH dehydrogenase subunit 3 DNA sequences. *Canadian Journal of Fisheries and Aquatic Science, 53*, 1165-1176.

Morita, K. and Yamamoto, S. (2002). Effects of habitat fragmentation by damming on the persistence of stream-dwelling charr populations. *Conservation Biology,* 16, 1318-1323.

Neville, H., Isaak, D., Thurow, R., Dunham, J. and Rieman, B. (2007). Microsatellite variation reveals weak genetic structure and retention of genetic variability in threatened Chinook salmon (*Oncorhynchus tshawytscha*) within a Snake River watershed. *Conservation Genetics,* 8, 133-147.

Nielsen, E. E., Hansen, M. M. and Loeschcke, V. (1999). Genetic variation in time and space: microsatellite analysis of extinct and extant populations of Atlantic salmon. *Evolution,* 53, 261-268.

Okazaki, T. (1986). Genetic variation and population structure in Masu salmon*Oncorhynchus masou* of Japan. *Bulletin of the Japanese Society of Scienctific Fisheries*, 52, 1365-1376.

Okazaki, T. (1990). Population structure of masu salmon, *Oncorhynchus masou*, in the species of the genus*Oncorhynchus*. Bulletin of the Institute of Zoology, *Academia Sinica,* 29, 17-25.

Oohara, I., Sawano, K. and Okazaki, T. (1997). Mitochondrial DNA sequence analysis of the masu salmon-phylogeny in the genus*Oncorhynchus. Molecular Phylogenetics and Evolution, 7*, 71-78.

Osinov, A. G. and Lebedev, V. S. (2000). Genetic divergence and phylogeny of the Salmoninae based on allozyme data. Journal of Fish Biology, 57, 354-381.

Peake, S., McKinley, R. S. and Scruton, D. A. (1997). Swimming performance of various freshwater Newfoundland salmonids relative to habitat selection and fishway design. *Journal of Fish Biology,* 51, 710-723.

Polyakova, N. E., Semina, A. V. and Brykov, V. A. (2006). The variability in chum salmon*Oncorhynchus keta* (Walbaum) mitochondrial DNA and its connection with the paleogeological events in the northwest Pacific. *Russian Journal of Genetics,* 42, 1164-1171.

Primmer, C. R., Veselov, A. J., Zubchenko, A., Poututkin, A., Bakhmet, I. and Koskinen, M. T. (2006). Isolation by distance within a river system: genetic population structuring of Atlantic salmon, *Salmo salar*, in tributaries of the Varzuga River in northwest Russia. *Molecular Ecology,* 15, 653-666.

Quinn, T. P. (1993). A review of homing and straying of wild and hatchery-produced salmon. *Fisheries Research,* 18, 29-44.

Quinn, T. P. and Dittman, A. H. (1990). Pacific salmon migrations and homing - mechanisms and adaptive significance. *Trends in Ecology and Evolution,* 5, 174-177.

Rogers, A. R. and Harpending, H. (1992). Population growth makes waves in the distribution of pairwise genetic differences. *Molecular Biology and Evolution,* 9, 552-569.

Sakata, K., Kondou, T., Takeshita, K., Nakazono, A. and Kimura, S. (2005). Movement of the fluvial form of masu salmon, *Oncorhynchus masou masou*, in a mountain stream in Kyushu, Japan. *Fisheries Science,* 71, 333-341.

Saitou, N. and Nei, M. (1987). The neighbor-joining method: a new method for reconstructing phylogenetic trees. *Molecular Biology and Evolution,* 4, 406-425.

Salo, E. O. (1991). Life history of chum salmon (*Oncorhynchus keta*). In: C. Groot and L. Margolis (eds.), *Pacific Salmon Life Histories* (pp. 231-309). Vancouver, UBC Press.

Sato, S., Kojima, H., Ando, J., Ando, H., Wilmot, R. L., Seeb, L. W., Efremov, V., Leclair, L., Buchholz, W., Jin, D.-H., Urawa, S., Kaeriyama, M., Urano, A. and Abe, S. (2004). Genetic population structure of chum salmon in the Pacific Rim inferred from mitochondrial DNA sequence variation. *Environmental Biology of Fishes,* 69, 37-50.

Spidle, A. P., Schill, W. B., Lubinski, B. A. and King, T. L. (2001). Fine-scale population structure in Atlantic salmon from Maine's Penobscot River drainage. Conservation Genetics, 2, 11-24.

Steele, C., Baumsteiger, J. and Storfer, A. (2009). Influence of life-history variation on the genetic structure of two sympatric salamander taxa. *Molecular Ecology,* 18, 1629-1639.

Sugiwaka, K.-i. and Kojima, H. (1984). Influence of individual density on smoltification in wild juvenile masu salmon (*Oncorhynchus masou*) in the Atsuta River. *Science Reports of Hokkaido Fish Hatchery,* 39, 19-37.

Suzuki, K.-i. T., Kobayashi, T., Matsuishi, T. and Numachi, K.-i. (2000). Genetic variability of masu salmon in Hokkaido, by restriction fragment length polymorphism analysis of mitochondrial DNA. *Nippon Suisan Gakkaishi,* 66, 639-646.

Tajima, F. (1989). Statistical method for testing the nautral mutation hypothesis by DNA polymorphism. *Genetics,* 123, 585-595.

Taylor, E. B. and Foote, C. J. (1991). Critical swimming velocities of juvenile sockeye salmon and kokanee, the anadromous and non-anadromous forms of *Oncorhynchus nerka*. *Journal of Fish Biology,* 38, 407-419.

Tonteri, A., Veselov, A. J., Titov, S., Lumme, J. and Primmer, C. R. (2007). The effect of migratory behaviour on genetic diversity and population divergence: a comparison of anadromous and freshwater Atlantic salmon*Salmo salar*. *Journal of Fish Biology,* 70, 381-398.

Utter, F. (2001). Patterns of subspecific anthropogenic introgression in two salmonid genera. *Reviews in Fish Biology and Fisheries,* 10, 265-279.

Vähä, J.-P., Erkinaro, J., Niemelä, E. and Primmer, C. R. (2007). Life-history and habitat features influence the within-river genetic structure of Atlantic salmon. *Molecular Ecology,* 16, 2638-2654.

Weber, E. D. and Fausch, K. D. (2003). Interactions between hatchery and wild salmonids in streams: differences in biology and evidence for competition. *Canadian Journal of Fisheries and Aquatic Sciences,* 60, 1018-1036.

Yamamoto, S., Morita, K., Koizumi, I. and Maekawa, K. (2004). Genetic differentiation of white-spotted charr (*Salvelinus leucomaenis*) populations after habitat fragmentation: spatial-temporal changes in gene frequencies. *Conservation Genetics,* 5, 529-538.

Yamamoto, T. and Edo, K. (2006). Factors influencing the breeding activity in male masu salmon, *Oncorhynchus masou*: the relationship with body size. *Environmental Biology of Fishes,* 75, 375-383.

Yu, J. N., Azuma, N., Yoon, M., Brykov, V., Urawa, S., Nagata, M., Jin, D. H. and Abe, S. (2010). Population Genetic Structure and Phylogeography of Masu Salmon (*Oncorhynchus masou masou*) Inferred from Mitochondrial and Microsatellite DNA Analyses. *Zoological Science,* 27, 375-385.

Zardoya, R., Castilho, R., Grande, C., Favre-Krey, L., Caetano, S., Marcato, S., Krey, G. and Patarnello, T. (2004). Differential population structuring of two closely related fish species, the mackerel (*Scomber scombrus*) and the chub mackerel (*Scomber japonicus*), in the Mediterranean Sea. *Molecular Ecology,* 13, 1785-1798.

In: Fish Ecology
Editor: Sean P. Dempsey
ISBN 978-1-61324-282-7

Chapter 5

FACULTATIVE CATADROMY IN THE FRESHWATER EEL GENUS *ANGUILLA* BETWEEN FRESH WATER AND SEA WATER HABITATS

Naoko Chino** *and Takaomi Arai
Atmosphere and Ocean Research Institute, The University of Tokyo
2-106-1, Akahama, Otsuchi, Iwate 028-1102, Japan

ABSTRACT

The freshwater eel of the genus*Anguilla*, being catadromous, migrate between fresh watergrowth habitats and offshore spawning areas. However, a number of recent studies found that thetemperate species*Anguilla anguilla*, *A. rostrata*, *A. japonica*, *A. australis* and *A. dieffenbachii*have never migrated into fresh water, spending their entire life in the ocean. Furthermore, those studies found an intermediate type between marine and freshwater residents, which appear to frequently move between different environments during their growth phase. The discovery of marine and brackish water residents suggests that anguillid eels do not all have to be catadromous, and it calls into question the generalized classification of diadromous fish. However, there has beenlittle available information concerning migration in tropical eels. In *A. marmorata,* showed three fluctuation patterns; (1) freshwater residence, (2) continuous residence in brackish water, and (3) residence in fresh water after recruitment, while returning to brackish water. Such migratory histories were found in other tropical eels, *A. bicolor bicolor* and *A. bicolor pacifica*. The*A. bicolor bicolor,* collected in a coastal lagoon of Indonesia,showed two patterns of habitat use, (1) constantly living in either brackish or sea waters with no fresh water life, and (2) habitat shift from fresh water to brackish or sea waters. The wide range of environmental habitat use indicated that migratory behavior of the tropical eel was facultative among fresh, brackish and marine waters during their growth phases after recruitment to the coastal areas. Further, the migratory behaviors of tropical eels appear to differ in each habitat in response to inter- and intra- specific competition. The results suggest that tropical eels have a flexible pattern of migration, with an ability to adapt to various habitats and salinities. This flexible habitat use was the same as that of temperate

*Corresponding author: N. Chino, E-mail: chino@aori.u-tokyo.ac.jp

eels. Thus, the migrations of anguillid eels into fresh water is clearly not an obligatory behavior.

Introduction

Fish that migrate between fresh water and sea water are called diadromous. The migration patterns of such species differ and have seasonal and life cycle variations. Only one percent of all fish in the world are diadromous (Myers 1949). Crossing the important habitat barrier from fresh water to sea water or vice versa during migration requires major physiological changes for survival. Diadromous fish have been subdivided into three categories (McDowall 1988);catadromous (migration into the sea for reproduction), anadromous (return from ocean to fresh water for breeding), and amphidromous (migration between sea and fresh water for feeding). Many diadromous species support important commercial and/or recreational fisheries. Thus, information regarding diadromous migrations would provide basic knowledge for both fish migration studies and fishery management, allowing effective and sustainable use of fish resources.

The freshwater eel of the genus*Anguilla*, being catadromous, migrates between fresh water growth habitats and offshore spawning areas. Fifteen species of *Anguilla* have been reported world-wide, ten of which are known to occur in tropical regions (Ege 1939). Of the latter, seven species/subspecies occur in the western Pacific around Indonesia, i.e. *A. celebesensis*, *A. interioris*, *A. nebulosa nebulosa*, *A. marmorata*, *A. borneensis*, *A. bicolor bicolor* and *A. bicolor pacifica* (Ege 1939, Castle and Williamson 1974, Arai et al. 1999). *A. borneensis* from Borneo Island was closest to the ancestral form among the 15 presently known species.The tropical species seem to be more closely related to the ancestral form than their temperate counterparts. Thus, studying the life history and migration of tropical eels may provide some clues to understanding the nature of primitive forms of catadromous migration in anguillid eels and how the migration of the species became established. The results may also contribute to understanding the evolutionary pathway of migration in the species as well as other diadromous fish species that migrate between fresh water and sea water habitats.

The life cycle of the freshwater eel has five principal stages: the leptocephalus, glass eel, elver, yellow eel and silver eel stages. The larvae, leptocephali, drift on ocean current and are transported by the current. The leptocephali leave oceanic currents after metamorphosing into glass eels and then typically migrate upstream as elvers four to eight months after hatching (Arai et al. 2001) to grow in the fresh waterhabitats during the yellow stage (immature stage). After upstream migration in the eel, the elvers become yellow eels and live in the fresh water habitats such as rivers and lakes. Then during the silver eel stage (early maturing stage) in autumn and winter, their gonads begin maturing and they start their downstream migration into the ocean and back out to the spawning area, where they spawn and die.

The migratory history of several species of anguillid eels have been studied using microchemical techniques that determine the ratios of strontium to calcium (Sr:Ca ratio) in their otoliths. The Sr:Ca ratio in the otoliths of fishes differs according to the time they spend in fresh waterand sea water; this has also been found to be true for anguillid eels (Arai et al., 2003a, b, 2004, 2006, 2008, 2009; Kotake et al., 2003, 2005, Chino et al. 2008, Chino and Arai 2009, 2010a, b, c). Early studies on the strontium level in eel otoliths of *Anguilla japonica* showed that the Sr:Ca level in their otolith strongly correlated with the salinity of the water and were little affected by other factors such as water temperature, food and physiological factors (Kawakami et al., 1998). Thus, the Sr:Ca ratios of otoliths could help in determining whether or not individual eels actually enter fresh water at the elver stage and remain in fresh water, estuarine or marine environments until the silver eel stage, or whether they move between different habitats with differing salinity regimes.

Otolith microchemistry studies have revealed that some yellow and silver eels of temperate anguillid eels never migrate into fresh water, but spend their entire life in the ocean. Application of otolith Sr:Ca ratios to trace the migratory history of eels also revealed otolith signatures intermediate to those of marine and freshwater residents of *A. anguilla* (Arai et al., 2006), *A. japonica* (Arai et al., 2003a, b, 2008; Kotake et al., 2003, 2005, Chino et al. 2008, Chino and Arai 2009), *A. rostrata* (Thibault et al., 2007), *A. australis*, and *A. dieffenbachii* (Arai et al., 2004), *A. marmorata* (Chino and Arai 2010a) and *A. bicolor bicolor* (Chino and Arai 2010b, c), all of which appeared to reflect estuarine residence, or showed clear evidence of switching between different salinity environments. It thus appears that a proportion of eels move frequently between different environments during their growth phase. Therefore, because individuals of several anguillid species have been found to remain in estuarine or marine habitats, it appears that anguillid eels do not all enter into fresh water environments and that these species display more a facultative catadromy (Chino and Arai, 2009, 2010b, c).

It is not clear why during the growth phase some eels migrate to fresh water and others do not. The occurrence of fish migration is generally explained by the existence of a difference in food abundance between marine and fresh water habitats, and Gross (1987) proposed that diadromy occurs when the gain in fitness from using a second habitat minus the migration costs of moving between habitats exceeds the fitness from staying in only one habitat. Juvenile anadromous salmon utilize fresh water habitats at high latitudes with low productivity, and they migrate to higher productivity habitats in the ocean for growth before returning to fresh water for breeding. In contrast, catadromous freshwater eels that recruit at low latitudes might migrate upstream into fresh water habitats of higher productivity for growth before returning to the ocean for breeding.

In this chapter, we synthesize the understandings for diversity of habitat preference and movements between fresh water and sea water habitatsby the *Anguilla* species. Further, we discuss the evolution and occurrence of diverse migration of eels based on recent studies.

Materials and Methods

This chapter synthesizes information following our publications: Arai et al. (2003a, b, 2008, 2009), Chino et al. (2008) and Chino and Arai (2009) for *Anguilla japonica*, Arai et al. (2004) for *A. australis*and *A. dieffenbachii*, *A. anguilla* for Arai et al. (2006), Chino and Arai (2010a) for *A. marmorata* and Chino and Arai (2010b, c) for *A. bicolor bicolor* (Figure 1).

In all studies, sagittal otoliths were extracted, and the otoliths were embedded in epoxy resin (Struers, Epofix). These otoliths were then ground to expose the core along the anterior-posterior direction in the frontal plane, using a grinding machine equipped with a diamond cup-wheel (Struers, Discoplan-TS), and polished further with an oxide polishing suspension on an automated polishing wheel (Struers, PdM-Force-20). Finally, they were cleaned using distilled water and ethanol, and dried at 50°C in an oven prior to examination. The ground surfaces of the otoliths were examined at 200 x with a light microscope, and photographs were taken to measure the "radius" of the elver mark (the distance from the otolith core to the elver check).

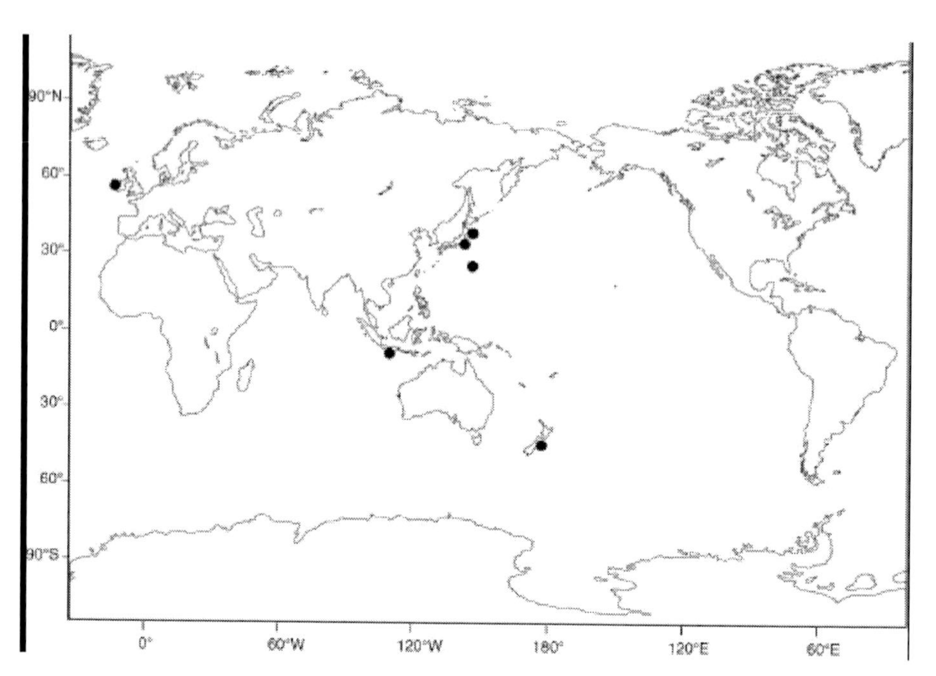

Figure 1.Study sites show the geographical range of each species used in this chapter. The source of information is Arai et al. (2003a, b, 2008, 2009), Chino et al. (2008) and Chino and Arai (2009) for *Anguilla japonica*from Japan, Arai et al. (2004) for *A. australis*and *A. dieffenbachii* from New Zealand, *A. anguilla* for Arai et al. (2006) from Ireland, Chino and Arai (2010a) for *A. marmorata* from Japan and Chino and Arai (2010b, c) for *A. bicolor bicolor* from Indonesia.

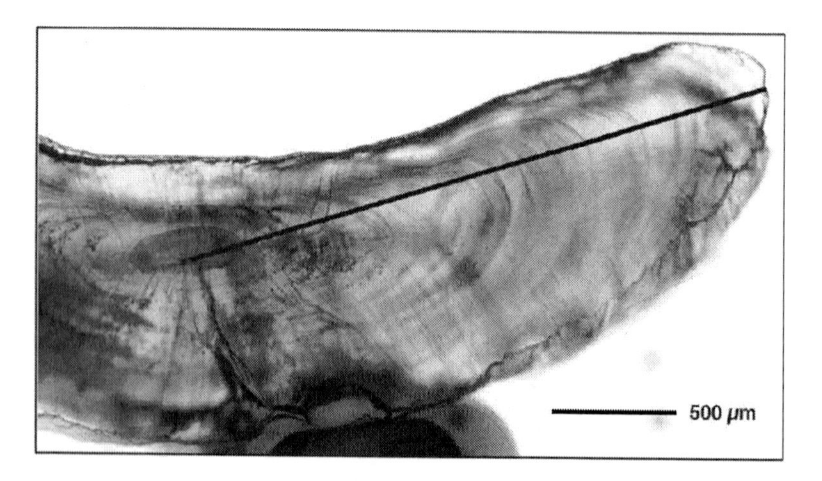

Figure 2.Image of life-history transect analyses of Sr and Ca concentrations in the freshwater eel, which were measured along a line down the longest axis of otolith from the core to the edge (black line) using a wavelength dispersive X-ray electron microprobe.

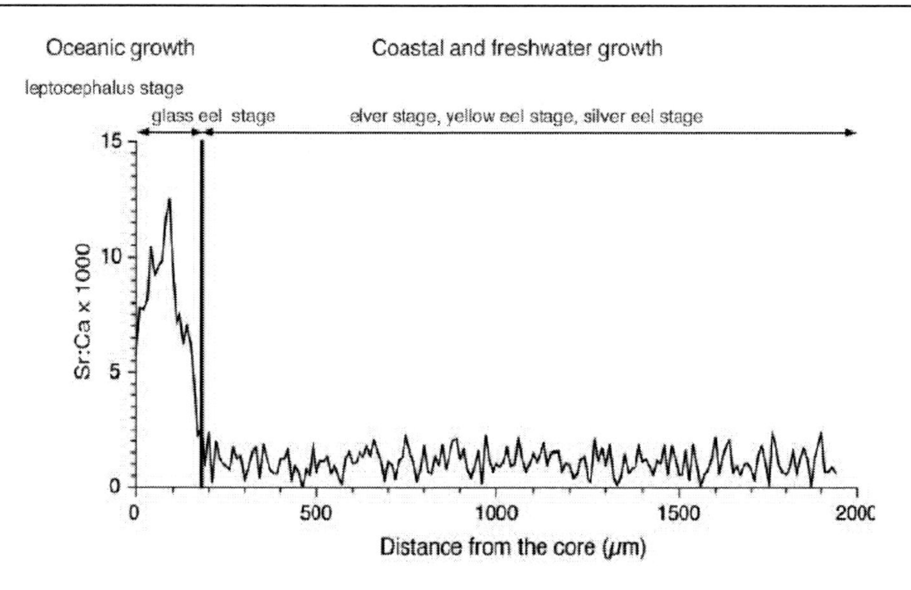

Figure 3. Fluctuation of otolith Sr:Ca ratios along a transect line from the core to the edge of the otolith with reference to growth stages and habitat uses in the life history of the freshwater eel.

For otolith microchemical analyses, all otoliths were Pt-Pd coated by a high vacuum evaporator. Otoliths from all specimens were used for life-history transect analyses of Sr and Ca concentrations, which were measured along a line down the longest axis of each otolith from the core to the edge (Figure 2) using a wavelength dispersive X-ray electron microprobe (JEOL JXA-8900R). Wollastonite ($CaSiO_3$) and Tausonite ($SrTiO_3$) were used as standards. The accelerating voltage and beam current were 15 kV and 1.2 x 10^{-8}A, respectively. The electron beam was focused on a point 10 μm in diameter, with measurements spaced at 10 μm intervals.

We calculated the average Sr:Ca ratios for the values outside the elver mark. Following the criteria of Chino and Arai (2010a), these specimens were categorized into "marine resident" (Sr:Ca $\geq$ 6.0 x 10^{-3}), "estuarine resident" (2.0 x 10^{-3} $\leq$ Sr:Ca < 6.0 x 10^{-3}) and "freshwater resident" (Sr:Ca < 2.0 x 10^{-3}).

RESULTS

Migratory History

The Sr:Ca ratios in the transects along the radius of each otolith showed the same common feature in all specimens (Figure 3). All otoliths had a common peak of high values of Sr:Ca ratios at the center of the otolith inside the elver mark (ca 150 μm), which roughly corresponded to the leptocephalus and early glass eel stages during their oceanic life (Arai et al. 1997). Outside of the high Sr core, there was a great variation in the change of the Sr:Ca ratios in the otoliths of eels from different habitats. The high Sr content in the central core regionduring the leptocephalus stage may derive from thelarge amounts of gelatinous extracellular matrix thatfills the body until metamorphosis. This material iscomposed of sulfated glycosaminoglycans (GAG),which are converted into other compounds during metamorphosis (Pfeiler 1984). The drastic decrease inSr at the outer otolith region in both river and sea water samples after metamorphosis to glass eels may occurbecause these sulfated polysaccharides have an affinityto alkali earth elements, and are particularly high inSr, suggesting that a high Sr content in the body has asignificant influence on otolith Sr content through the saccularepithelium in the inner ear, and the sudden loss of Sr-rich GAG during metamorphosis probably results in the lower Sr concentration in otoliths after metamorphosis (Otake et al. 1994 Arai et al. 1997).

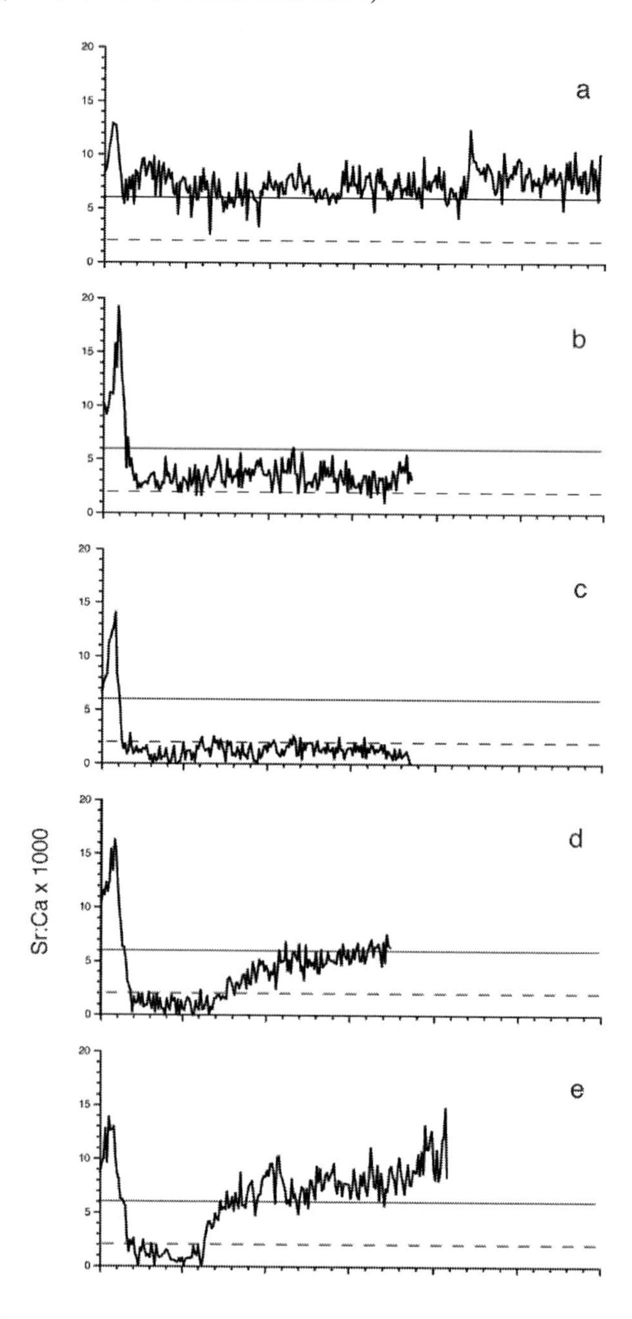

Figure 4. (continued).

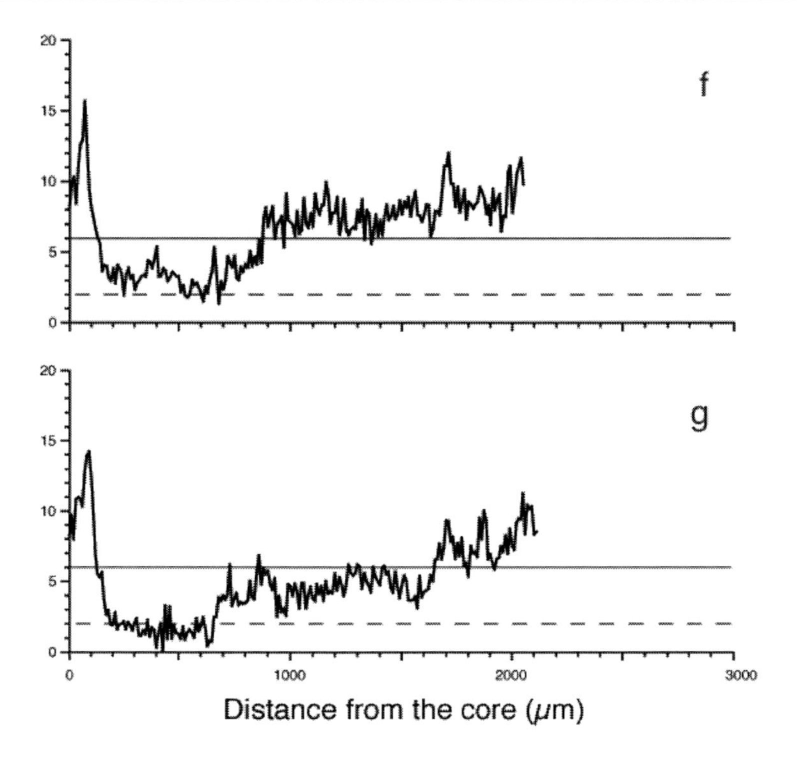

Figure 4. Plots of the otolith Sr:Ca ratios along transect lines from the core to the edge of the otolith for seven representative specimens of the freshwater eel.Eels exhibiting a constant type lived in either (a)marine water, (b) brackish wateror (c) fresh water, throughout their lives. Eels exhibiting a switch type moved among habitats. The varieties included eels that moved from (d) fresh water to brackish water, (e) fresh water to marine water, (f); brackish water to marine water, (g) multiple movements among fresh water, brackish water, and marine water. The solid line in each panel indicates the marine water life period ($\geq 6.0 \times 10^{-3}$ in Sr:Ca ratios), and the dotted line in each panel indicates the fresh water life period ($< 2.0 \times 10^{-3}$ in Sr:Ca ratios).

Outside of the high Sr:Ca core, there was great variation in the Sr:Ca ratios in the otoliths of eels from different species and habitats. The migratory patterns revealing the life history were generally classified into two types (Figure 4). The first pattern can be labeled the constant type, with constant Sr:Ca ratios along the life history transect, which suggests that these specimens lived in the same environmentin either fresh water, brackish water or sea water throughout their lives (Figure 4a, b , c). The change in Sr:Ca values outside the elver mark in the constant type was generally divided into three types (Figure 4a, b , c): (1) relatively high values of more than 6.00×10^{-3} with no movement into fresh water (Figure 4a), (2) intermediate values of 2.00-6.00 x 10^{-3}(Figure 4b), and (3) constantly low values less than2.00 x 10^{-3}(Figure 4c).The second migratory pattern can be labeled the switch type (Figure 4d, e, f, g) with Sr:Ca ratios shifting between two phases from a low phase (less than2.00 x 10^{-3}) to a middle phase (range: 2.00-6.00 x 10^{-3}) (Figure 3d), from a low phase to a high phase (more than 6.0×10^{-3}) (Figure 4e), or from a middle phase to a high phase (Figure 4f), which suggests that these specimens shifted their resident environment. Further, specimens showed multiple shifts, with several low and high Sr:Ca ratio phases along the life history transect, which suggests that these specimens moved to different environments several times in their lives (Figure 4g).

Index of Habitat Use

In order to quantitatively estimate the general habitat use of each specimen based on its mean Sr:Ca ratio values,we calculated an index for thedegree of sea waterresidence as follows.Since all specimens had experiencedthe same common marinelife as a preleptocephalus, leptocephalus,metamorphosing larva andearly glass eel during their long tripfrom the spawning area to coastal waters, the values of Sr:Ca inside theelver mark could be excluded from each life-historytransect (see Figure 3) when estimating the degree of sea water residence for the juvenile stages of each individualafter arrival in coastal waters at the early glass eel stage. The mean Sr:Ca ratio values outside the elver mark in each eel indicated that the habitat use was variable with interspecific variationsafter their recruitment to the coastal waters as glass eels (Table 1).

Table.1 Habitat use as an indication of ranging of otolith Sr:Ca ratios for tropical and temperate eels of genus *Anguilla*

Species	Country	Sr/Ca ratios	Reference
Tropical eels			
A. bicolor bicolor	Indonesia	1.71 - 7.40 x 10^{-3}	Chino and Arai 2010b, c
	Philippines	2.65 ± 1.20 x 10^{-3}	Briones et al. 2007
A. bicolor pacifica		4.70 ± 1.68 x 10^{-3}	
		1.53 ± 0.92 x 10^{-3}	
A. marmorata	Japan	1.1 - 3.2 x 10^{-3}	Chino and Arai 2010a
	Taiwan	1.6 - 2.2 x 10^{-3}	Shiao et al. 2003
Temperate eels			
A. anguilla	Ireland	1.0 - 8.7 x 10^{-3}	Arai et al. 2006
	Sweden	0.7 - 7.5 x 10^{-3}	Tzeng et al. 2000
A. japonica	Japan	1.0 - 8.5 x 10^{-3}	Kotake et al. 2005
	Taiwan	1.8 - 5.5 x 10^{-3}	Shiao et al. 2003
A. rostrata	Canada	0.38 - 7.46 x 10^{-3}	Jessop et al. 2002
A. australis	New Zealand	1.83 - 7.35 x 10^{-3}	Arai et al. 2004
A. dieffenbachii		1.82 - 4.02 x 10^{-3}	

DISCUSSION

Diverse Migration between Fresh Water and Sea Water habitats

The most significant finding revealed in the otolith microchemical analyses was thatthe occurrence of marine resident eels that had nevermigrated into fresh waterhabitat was confirmed in many species and localities in coastal waters for both temperate and tropical eels. Another significant result wasthe finding of intermediate constant type of eel migration (estuarine resident) and switch type of the migration (Figure 4). A predominance of eels that move once or several times between habitats of different salinity has been reported for *A. rostrata* (Thibault et al. 2007) and other localities (Tzeng et al. 2000, Shiao et al. 2003, Daverat et al. 2006) (Table 1). These findings strongly suggestthat the freshwater eel *Anguilla*has a flexible migrationstrategy with a high degree of behavioral plasticity andan ability to utilize the full range of salinity as juveniles. Further, the results demonstrated that the eels found in coastal and estuarine habitats can be residentin these areas and also may move back and forthbetween fresh water and sea water (Figure 5). Thus, theclassification of anguillid eels as being catadromous and having a fresh water growth stage clearly needs revision, because it is nowevident that their movement into fresh water is not anobligate migratory pathway, and should be defined asa facultative catadromy, with ocean and estuarine residentsas ecophenotypes.

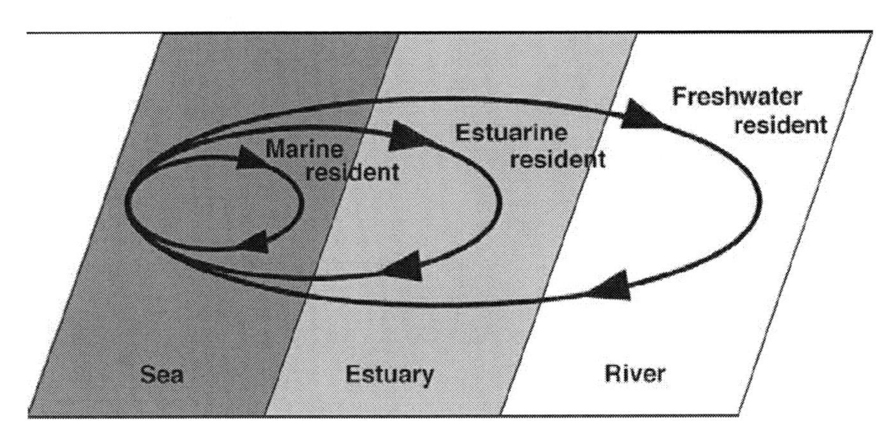

Figure 5. Diagrammatic presentation of the life cycle of the freshwater eel with special reference to their migratory histories. Freshwater resident eels migrate upstreamafter arriving at the estuary during the glass eel stage and stay in the freshwaterhabitat until their downstream spawning migration as silver-phase eels, while marine resident eels never migrate into the freshwater habitat and spend their entire life in theocean. An intermediate type of estuarine resident eels occurs mainly in the brackish waterhabitat in estuaries which may move back and forth between freshwater, brackishwater and/or sea water.

In the tropical region, several eel species are living sympatrically in a river system. Chino and Arai (2010b) found thatthe migratory patterns of *A. bicolor bicolor* were mainly of the switch type, moving from fresh waterto sea water environments (75%). The other eels resided consistently in either sea water (20%) or brackish water (5%). There were no eelsthat showed constant freshwaterresidence in Segara Anakan, central Java, Indonesia. In the Philippines, some *A. bicolor bicolor* eels showed freshwater residence (37%), while mostshowed estuarineresidence (63%) (Briones et al. 2007). Thus, *A. bicolor bicolor* in the Philippines preferred to live in a brackish water environment. However, this finding is quite different from that of Indonesia, which showed only a 5% estuarineresidence in Segara Anakan. In Segara Anakan, there isa wide range of salinity environments in which eels may live. Interspecific competition is one of the reasons for the dissimilarity of habitat use in this species. In central Java there were three species, *A. bicolor bicolor*, *A. marmorata* and *A.nebulosa nebulosa*, and in the Philippines, three species were also found, *A. bicolor bicolor*, *A. bicolor pacifica* and *A. marmorata*. Chino and Arai (2010b)discovered*A. bicolor bicolor* in either brackish water or sea water, although they did not collect eels from the upstream area of the river where there was no influence of a tidal effect. In the Philippines, *A. bicolor bicolor* also occurred in the downstream area of the river (Briones et al. 2007). Therefore, interspecific competition may occur between (among) eel species ateach site, and each species has a restricted distribution in either fresh water, brackish water or sea water habitats. In *A. marmorata*, the habitat preference in terms of salinity environment may depend on the habitat where there are either multiple or mono species of anguillid eels. In the Philippines and Taiwan, *A. marmorata* tended to live in fresh water environments (Shiao et al. 2003, Briones et al. 2007), while those in Ogasawara Island of Japanlived in brackishwater environments (Chino and Arai 2010a). There were*A. bicolor bicolor*, *A. bicolor pacifica* and *A. marmorata* in the Philippines and *A. marmorata* and *A. japonica* in Taiwan, while only *A. marmorata*was observed in Ogasawara Island. In Oagsawara Island, *A. marmorata* could live in every environment with no interspecific competition, and thus estuarine-dependent eels may be more abundant than in the Philippines and Taiwan. The results suggest that the determination of the distribution of eels at each site may occur by interspecific competition.

Cause of Occurrence of Migratory Plasticity

Fish migration is generally explained by a differencein food abundance between marine and fresh water habitats (Gross 1987). Juvenile anadromous salmonutilize low-productivity fresh water habitats at highlatitudes, and they migrate to higher-productivityhabitats in the ocean for growth before returning to fresh water for breeding. In contrast, catadromousfreshwater eels that recruit at low latitudes migrateupstream into fresh water habitats for higher productivity of growth before returning to the ocean for breeding.Therefore, a latitudinal cline might be predicted inwhich marine resident freshwater eels would occurmore frequently at higher latitudes where the productivityof the fresh water habitat is lower compared to the ocean. However, the habitat use as an index of otolith Sr:Ca ratios of the tropical eel *A. bicolor bicolor* suggested that the species had more flexible migratory behaviors in ambient waters than did other tropical eels such as *A. marmorata*. (Table 1). The habitat preference in *A. bicolor bicolor* during the growth phase was the same as that of the temperate eels *A. anguilla*, *A. rostrata*, *A. japonica*, *A. australis* and *A. dieffenbachii* (Table 1). Tsukamoto et al. (2002) hypothesized that higher proportions of brackish water- and marine-dependent eels live at the higher latitudes than in the subtropical to tropical regions. Furthermore, Daverat et al. (2006) assumed that eels found at higher latitudes exhibited a greater probability of remaining in the lower reaches of watersheds in brackish water based on previously published results of the temperate eels *A. anguilla*, *A. rostara* and *A. japonica*. In the St. Jean estuary, Canada (48°N), 60% of *A. rostrata* remained in the estuary compared to 35% estuarine dependence in the Hudson River estuary, United States (42°N). Most of *A. anguilla* were found to be estuarine residents in the Baltic Sea, Sweden (55 to 60°N) while 44% of eels were resident in Gironde estuary, France (44°N). In *A. japonica*, 5% of the eels showed marine residence in the Koaping River, Taiwan (22°N) compared to 24% in Mikawa Bay, Japan (34°N). However, the present and previous results related to the habitat of tropical eels did not support these ideas. We only found either brackish water- or marine-dependent eels without the typical fresh water-dependent eel of *A. bicolor bicolor* in Indonesia, in the equatorial region (7°S). Further, eels of most migration types were also found in either brackish water- or marine-dependent eels in *A. bicolor bicolor* and *A. bicolor pacifica* in Phillippines (18°N, Briones et al. 2007). The ability of anguillid eels to reside in environments of various salinities would be a common feature between tropical and temperate species without latitudinal cline. This evidence of geographical variability among eel species suggests that habitat use is determined by environmental conditions in each site.

In anadromous salmon, freshwater residents or landlockedpopulations that do not migrate to the oceanoften occur, especially near the southern limit of their geographical distribution (McDowall 1988). The ancestorsof salmon, which originated in fresh water,expanded their growth habitat into the ocean whiletheir breeding place remained in fresh water. Sincereproduction is physiologically costly, it was hypothesizedthat the migratory behavior remained conservativethrough its evolutionary process (Tcharnavin 1939). Freshwater eels of the genus*Anguilla* are consideredto have originated from a marine ancestor, and all anguilliformfishes except *Anguilla* are marine species; thus, themarine breeding habits of *Anguilla* are probably a conservativetrait. This suggests the hypothesis that atleast some species of catadromous eels have never lostthe ability to be resident in marine habitats during thejuvenile growth phase, but it is unknown whether thisis due to a remnant genetic trait that determines if anindividual will enter fresh water or not, or if it is simply due to behavioral plasticity that enables each speciesto use the maximum range of habits.

Another hypothesis for the occurrenceof marine resident anguillideels would be ecological competitionwith other species (Moriarty 1978).In the case of *A. anguilla*, theymay have strong competition withthe conger eel *Conger conger*, especiallyin regard to predation at lowerlatitudes, and thus there are few reportsof the occurrence of Europeaneels in the ocean from the Centraland South European and Mediterraneancoasts where conger eels areplentiful. In contrast, the North Seaand Baltic Sea have no conger eels,but European eels are abundant andcommercially exploited. Thus, such intra- and inter-specific competition might be important factors to consider regarding the migration plasticity in the freshwater eel.

Relative Contribution of Migratory Type on the Next Generation

It is not known if all populations display the same utilization of both estuarine and marine environments in addition to the typical fresh waterenvironments of the species. Furthermore, due to the difficulty of collecting silver eels in the open ocean, there is little information available on estimating the relative contribution of the migratory type to reproduction. To address this question, we examined the Sr:Ca ratios in the otoliths of silver eels of *A. japonica*that just began their spawning migration to the open ocean and were caught offshore on the Pacific side of Japan (Chino and Arai 2009).Three migratory types, which were categorized as freshwater resident, estuarine resident, and marine resident were found. The estuarine resident was dominant (59%), followed by marine resident(22%) and freshwater resident (19%).The low proportion of freshwater residents from the spawning migration season suggested that the estuarine and marineresident eels inhabiting the nearby coastal areas might make a larger reproductive contribution to the next generation in this area. The similarity of habitat use patterns were found in other species such as temperate eels *A. anguilla*, *A. rostrata*, *A. australis* and *A. dieffenbachii* and tropical eels *A. bicolor bicolor* and *A. marmorata* (Table 1), although not all samples were categorized as silvering stage. Thus, the estuarine and marineresident eels may make a substantialcontribution to the spawning stock, and it is a common feature in the anguillid eel. Freshwater, estuarine and marine resident eels begin their spawning migration toward the open ocean at about the same time. This type of synchronization of migration and gonadal maturation, and the apparent predominance of estuarine and marine habitats, has important implications for the conservation of this species. It implies that eels from both fresh water and marine habitats can mix together during the spawning migration and potentially contribute to the next generation, and that estuarine and marine habitats may be very important for the eels.

Interestingly, an introduced eel, *A. anguilla* at the silver stage was found in Tokyo Bay, Japan (Arai et al. 2009). The eel might have either escaped from a culture pond or been released into a river. The eel spent its entire life in a fresh water environment just before spawning migration, as was shown by the value of the Sr:Ca ratio, which showed a low level during the glass eel and yellow eel stages (Arai et al. 2009). Furthermore, the fact that *A. anguiila* was found in Japanese waters, and that the eel matured and began its spawning migration, suggests that the introduced eel could adapt and survive in a foreign environment. Other non-native eels have been found thriving in the Uono River, Niigata Prefecture (Miyai et al., 2004) and Mikawa Bay (Okamura et al., 2008), Japan. Although the number of introduced eels such as *A. anguilla* has recently decreased (Okamura et al., 2008) in Japanese waters, it is important to study the life history and behavior of eels introduced in waters to conserve the native eel population.

Conclusion

Regardless of species or site, both temperate eels *A. anguilla*, *A. rostrata*, *A. japonica*, *A. australis* and *A. dieffenbachii* and tropical eels *A. bicolor bicolor* and *A. marmorata* showed a wide range of habitat use patterns. Such diverse migration was the common feature in thefreshwater eel. Thus, the migrationof anguillid eels into fresh wateris clearly not an obligatory pathway. The evolutionary pathway of the diverse migration of anguillid eels is still unclear. However, it may be attributed to genetics or environmental adaptation as in other diadromous fish (Nordeng, 1983, Gross, 1985). Mitochondrial DNA (mtDNA) analysis could not find any genetic differences in *A. anguilla* (Daemen et al. 2001) and *A. japonica* (Sang et al., 1994), and thus, the population of eels is considered to be panmictic. There is a widely held view that life histories in animals are selected for and adapted to maximizing the production of progeny (Schaffer and Elson1975, Stearns1977, Dingle1980, Gross, 1985). The persistence of migration needs to be seen in relation to the balance of the advantages obtained from migration and the costs incurred by the population and/or species. These advantages include such aspects as increased food supply, avoidance of potentially harmful environmental conditions and/or movement to more favorable ones, the occupation of habitats that have specific or specialized habitat requirements, and the availability of more living space. In the freshwater eel *Anguilla*, either estuarine resident eels or marine resident eels make a larger reproductive contribution to the next generation. Thus, both estuarine and marine habitats are important for eels.

References

Arai, T., Otake, T. and Tsukamoto, K. (1997) Drastic changes in otolith microstructure and microchemistry accompanying the onset of metamorphosis in the Japanese eel *Anguilla japonica*. *Marine Ecology Progress Series* 161: 17-22.

Arai, T., Aoyama, J., Limbong, D. and Tsukamoto, K. (1999) Species composition and inshore migration of the tropical eels *Anguilla* spp. recruiting to the estuary of the Poigar River, Sulawesi Island.*Marine Ecology Progress Series* 188: 299-303.

Arai, T., Limbong, D., Otake, T. and Tsukamoto, K. (2001) Recruitment mechanisms of tropical eels, *Anguilla* spp., and implications for the evolution of oceanic migration in the genus*Anguilla*. *Marine Ecology Progress Series* 216: 253-264.

Arai, T., Kotake, A., Ohji, M., Miyazaki, N. and Tsukamoto, K. (2003a) Migratory history and habitat use of Japanese eel *Anguilla japonica* in the Sanriku Coast of Japan. *Fisheries Science* 69:813-818.

Arai, T., Kotake, A., Ohji, M., Miller, M. J., Tsukamoto, K. and Miyazaki, N. (2003b) Occurrence of sea eels of *Anguilla japonica* along the Sanriku Coast of Japan. *Ichthyological Research* 50: 78-81.

Arai, T., Kotake, A., Lokman, P. M., Miller, M. J. and Tsukamoto, K. (2004) Evidence of different habitat use by New Zealand freshwater eels, *Anguilla australis* and *A. dieffenbachii*, as revealed by otolith microchemistry. *Marine Ecology Progress Series* 266:213-225.

Arai, T., Kotake, A. andMcCarthy, T. K. (2006) Habitat use by the European eel *Anguilla anguilla* in Irish waters. Estuarine,*Coastal and Shelf Science* 67:569-578.

Arai, T., Kotake, A. and Ohji, M. (2008) Variation in migratory history of Japanese eels, *Anguilla japonica*, collected in the northernmost part of its distribution. *Journal of the Marine Biological Association of the United Kingdom*88: 1075-1080.

Arai, T., Chino, N. and Kotake, A. (2009)Occurrence of estuarine and sea eels *Anguilla japonica* and a migrating silver eel *Anguilla anguilla* in Tokyo Bay area, Japan. *Fisheries Science* 75: 1197-1203.

Briones, A. A., Yambot, A. V., Shiao, J. C., Iizuka, Y. andTzeng, W. N. (2007) Migratory pattern and habitat use of tropical eels *Anguilla* spp. (teleostei: Anguilliformes: Anguillidae) in the Philippines, as revealed by otolith microchemistry. *The Raffles Bulletin of Zoology* 14:141-149.

Castle, P. H. J. and Williamson, G. R. (1974) On the validity of the freshwater eel species*Anguilla ancestralis* Ege from Celebes. Copeia 2:569-570.

Chino, N., Yoshinaga, T., Hirai, A. andArai, T. (2008) Life history patterns of silver eels *Anguilla japonica* collected in the Sanriku Coast of Japan. *Coastal Marine Science* 32: 54-56.

Chino, N. and Arai, T. (2009) Relative contribution of migratory type on the reproduction of migrating silver eels, *Anguilla japonica*, collected off Shikoku Island, Japan. *Marine Biology* 156:661-668.

Chino, N. and Arai, T. (2010a) Migratory history of the giant mottled eel (*Anguilla marmorata*) in the Bonin Islands of Japan.*Ecology of Freshwater Fish* 19: 19-25.

Chino, N. and Arai, T. (2010b) Occurrence of marine resident tropical eel *Anguilla bicolor bicolor* in Indonesia. Marine Biology 157: 1075-1081.

Chino, N. and Arai, T. (2010c) Habitat use and habitat transitions in the tropical eel, *Anguilla bicolor bicolor*.*Environmental Biology of Fishes* 89: 571-578.

Daemen, E., Cross, T., Ollevier, F. and Volckaert, F. A. M., (2001) Analysis of the genetic structure of European eel (*Anguilla anguilla*) using microsatelliteDNA and mtDNA markers. *Marine Biology* 139: 755-764.

Daverat, F., Limberg, K. E., Thibault, I., Shiao, J. C., Dodson, J. J., Caron, F., Tzeng, W. N., Iizuka, Y. and Wickström, H. (2006) Phenotypic plasticity of habitat use by three temperate eel species, *Anguilla anguilla, A. japonica* and *A. rostrata. Marine Ecology Progress Series* 308:231-241.

Dingle, H. (1980) Ecology of juvenile grey mullet: a short review. *Aquaculture*19: 21-36.

Ege, V. (1939) A revision of the Genus *Anguilla* Shaw. Dana Report 16 (13):8-256.

Gross, M. R. (1985) Disruptive selection for alternative life histories in salmon. *Nature* 313: 47-48.

Gross, M. R. (1987) Evolution of diadromy in fishes. *American Fisheries Society Symposium* 1:14–25.

Kawakami, Y., Mochioka, N., Morishita, K., Toh, H. andNakazono, A. (1998) Determination of the freshwater mark in otoliths of Japanese eel elvers using microstructure and Sr/Ca ratios. *Environmental Biology of Fishes*53: 421-427.

Kotake, A., Arai, T., Ozawa, T., Nojima, S., Miller, M. J. and Tsukamoto, K. (2003) Variation in migratory history of Japanese eels, *Anguilla japonica*, collected in coastal waters of the Amakusa Islands, Japan, inferred from otolith Sr/Ca ratios. *Marine Biology* 142:849-854.

Kotake, A., Okamura, A., Yamada, Y., Utoh, T., Arai, T., Miller, M. J., Oka, H. P. and Tsukamoto, K. (2005) Seasonal variation in migratory history of the Japanese eel, *Anguilla japonica*, in Mikawa Bay, Japan. *Marine Ecology Progress Series* 293:213-221.

McDowall, R. M. (1988) *Diadromy in fishes.* Croom Helm, London.

Miyai, T., Aoyama, J., Sasai, S., Inoue, J. G., Miller, M. J. and Tsukamoto, K. (2004) Ecological aspects of the downstream migration of introduced European eels in the Uono River, Japan. *Environmental Biology of Fishes 71: 105-114.*

Moriarty, C. (1978) Eels, a natural and unnatural history. *Newton Abbot,*David and Charles.

Myers, G. S. (1949) Usage of anadromous, catadromous and allied terms for migratory fishes. *Copeia*, 1949: 89-97.

Nordeng, H. (1983) Solution to the char problem based on arctic char (*Salvelinus alpinus*) in Norway. *Canadian Journal of Fisheries and Aquatic Sciences 40*: 1372-1387.

Okamura, A., Zhang, H., Mikawa, N., Kotake, A., Yamada, Y., Utoh, T., Horie, N., Tanaka, S., Oka, H. P. and Tsukamoto, K. (2008) Decline in non-native freshwater eels in Japan: ecology and future perspectives. *Environmental Biology of Fishes* 81: 347-358.

Otake, T., Ishii, T., Nakahara, M. and Nakamura, R. (1994) Drastic changes in otolith strontium/calcium ratios in leptocephali and glass eels of Japanese eels Anguilla japonica. *Marine Ecology Progress Series* 112: 189-193.

Pfeiler, E., (1984) Glycosaminoglycan breakdown during metamorphosis of larval bone fish Albula. *Marine Biology Letters* 5: 241-249.

Sang, T., Chang, H., Chen, C. and Hui, C. (1994) Population structure of the Japanese eel, *Anguillajaponica. Molecular Biology and Evolution*11: 250-260.

Schaffer, W. N. and Elson, P. F. (1975) The adaptative significance of variations in life history among local populations of Atlantic salmon in North America. *Ecology*56: 577-590.

Secor, D. H. and Rooker, J. R. (2000) Is otolith strontium a useful scalar of life cycles in estuarine fishes? *Fisheries Research* 46:359-371.

Shiao, J.C., Iizuka, Y., Chang, C. W. and Tzeng, W. N. (2003) Disparities in habitat use and migratory behavior between tropical eel *Anguilla marmorata* and temperate eel *A. japonica* in four Taiwanese rivers. *Marine Ecology Progress Series*261:233-242

Stearns, S. C. (1977) The evolution of life history traits-a critique of the theory and a review of the data. *Annual Review of Ecology and Systematics 8:* 145-171.

Tcharnavin, V. (1939) The origin of salmon its ancestry marineor freshwater? *Salmon and Trout Magazine* 95:120–140.

Tesch, F. W. (2003) The Eel. *Biology and management of anguillid eels.* London,Chapman and Hall.

Thibault, I., Dodson, J. J., Caron, F., Tzeng, W. N., Iizuka, Y., Shiao, J. C.(2007) Facultative catadromy in American eels: testing the conditional strategy hypothesis. *Marine Ecology Progress Series*344: 219–229.

Tsukamoto, K., Aoyama, J. and Miller, M. J. (2002) Migration, speciation, and the evolution of diadromy in anguillid eels. *Canadian Journal of Fisheries and Aquatic Sciences*59:1989–1998.

Tzeng, W. N. (1996) Effects of salinity and ontogenic movements on strontium:calcium ratios in the otoliths of the Japanese eel, *Anguilla japonica* Temminck and Schlegel. *Journal of Experimental Marine Biology and Ecology 199: 111-122.*

Tzeng, W. N., Wang, C. H., Wickström, H. andReizenstein, M. (2000) Occurrence of the semi-catadromous European eel *Anguilla anguilla* in the Baltic Sea. *Marine Biology* 137:93-98.

In: Fish Ecology
Editor: Sean P. Dempsey

ISBN 978-1-61324-282-7

Chapter 6

IMPACTS OF ANTHROPIC FACTORS ON NATIVE FRESHWATER FISH IN BRAZILIAN SEMIARID REGION

Sathyabama Chellappa[a,*], Wallace Silva do Nascimento[b,†], Thiago Chellappa[c‡] and Naithirithi T. Chellappa[a]

[a]Post-Graduate Programme in Ecology, Department of Oceanography and Limnology, Center of Bioscience, Universidade Federal do Rio Grande do Norte (UFRN), Praia de Mãe Luiza, s/n, Natal, RN. Brazil

[b]Post-Graduate Programme in Psychobiology, Department of Physiology, Center of Bioscience, Universidade Federal do Rio Grande do Norte (UFRN), Campus Universitário, Rio Grande do Norte, Brazil

[c]Post-Graduate Programme in Material Science and Engineering, Laboratory of Catalysis and Petrochemistry, Universidade Federal do Rio Grande do Norte (UFRN), Av. Salgado Filho, 3000, Lagoa Nova, Natal, RN, Brazil

ABSTRACT

Native fish fauna of tropical semiarid freshwater ecosystems in Brazil are key representatives of the Neotropical region. Ichthyofauna of this region comprises of the orders Characiformes, Perciformes, Siluriformes and Synbranchiformes, besides invasive and exotic species. The population increase and the continued demand for quality fish generate the need for aquaculture as an alternative means of increasing fish production. Consequently, anthropic factors ranging from deforestation, construction of reservoirs on small rivers, introduction of invasive fish species to non-sustainable fishculture practice, can generate negative impacts on native species. Within these anthropic factors, proliferation of invasive and exotic fish species in the freshwater ecosystems has been deemed as a threat to the integrity of native species. However, the extent to which these factors can interfere and modify semiarid freshwater ecosystems remains an open

* To whom correspondence should be addressed. E-mail address: chellappa.sathyabama63@gmail.com; ntchellappa@ufrnet.br

† E-mail address: wallacesnbio@hotmail.com

‡ E-mail address: thiagochellappa@yahoo.com.br

question. This chapter focuses on how anthropic factors directly impact on the native fish species in semiarid freshwater ecosystems. Conversions of vegetation rich rural lands to urban areas degrade streams by qualitatively altering the composition, structure and function of their aquatic ecosystems. Construction of reservoirs can undoubtedly alter the natural hydrological regime of water bodies located downstream, and adversely impact on the migratory fish of this region. Introduction of exotic and carnivorous fish species from other hydrographical basins has significantly mediated to the decline of the native fish species, due to their competitive ability to thrive in semiarid aquatic ecosystems. Deforestation degrades riverine habitats and reduces ecological diversity of fish, besides causing pluvial shifts. Tropical semiarid Brazil is characterized with infrequent spells of rain interspersed with prolonged dry season. In tropics, rainfall is the main environmental driver that modulates the timing of fish spawning period. Changes in these environmental constraints present them with new challenges to survival and reproductive success. Since freshwater fish is important as a source of protein and provides sustained revenue, declining fish populations can have far reaching socioeconomic impacts which are compelling reasons for proper management. Strict regulations on introductions of fish species and restrictions on release of pollutants to aquatic ecosystems would reverse ecosystem degradation, improve water quality and conserve fish diversity.

Keywords: Tropical semiarid region, native freshwater fish species, anthropogenic impacts.

Chapter Highlights

Studies concerning impacts of anthropic activities on semiarid freshwater fish in Brazil have not yet been reported. The following questions were addressed:

1. What is the composition and abundance of native, non native and exotic fish communities in the Brazilian tropical semiarid freshwater ecosystems?
2. What are the principal anthropic impacts which negatively influence the fish in the semiarid freshwater ecosystems?
3. What mitigations would best preserve the fish species of these habitats?

This information could contribute to a better management of aquatic environments and the conservation of the tropical semiarid freshwater native fish communities.

INTRODUCTION

Fish represent the largest component of global vertebrate diversity, and are more diverse at all taxonomic levels, consisting of 57 orders of living fishes, with 482 families and more than 25,000 species. Out of this, more than 10,000 fish species occur in freshwater ecosystems (Nelson, 1994; Matthews, 1998). Fish is particularly important for the diet and nutrition of the rural population, as it is one of the affordable forms of animal protein. Introduction of non native species is among the most important, least controlled and reversible on the world's aquatic ecosystems, strongly affecting their biodiversity, biogeochemistry and economic uses. This is one of the dominant outcomes of human activities (Lockwood et al. 2007). There is great need for understanding the anthropogenic damages caused to freshwater fish and aquatic ecosystems for saving these for the future. Although studies on anthropogenic damages caused to freshwater fish are backed up by considerable literature support from world over, there is a paucity of information related to such studies in the tropical semiarid region of Brazil.

The Neotropical Region is one of the six major zoogeographical regions of the world which extends from Mexico to the southern most tip of South America. Brazil is an important part of this region which holds the largest biodiversity on the planet, accounting for around 13% of the total world biota (Queiroz et al., 2006). Although most of the country is located within the tropical zone, the climate of Brazil varies considerably from the Northern tropical zone to the temperate zones below the Tropic of Capricorn. Thus Brazil exhibits the equatorial, tropical, semi-arid, highland tropical and subtropical climatic regions. Semiarid climate generally describes regions that receive low annual rainfall (250-500 mm), and the semiarid region of Northeastern Brazil, considered as the area of drought polygon, has distinctive scrub vegetation referred to as "Caatinga", consisting of xerophytic low thorny bushes adapted to the semiarid evapotranspritation driven climate, which covers over 10% of the Brazilian territory with twenty million inhabitants. It is located between 3°S 45°W and 17°S 35°W, extending across eight administrative states of Brazil: Piauí, Ceará, Rio Grande do Norte,Paraiba, Pernambuco, Alagoas, Sergipe, Bahia,and parts of Minas Gerais (Andrade-Lima, 1981; Leal et al., 2003). The Caatinga is considered to be one of the 37 Wilderness Areas of the World, and it plays an important role in the maintenance of regional macro-ecological process, as well as indirectly supporting regions with diversity and endemism in Brazil (Leite and Machado, 2010).

The semiarid Caatinga biome of Northeastern Brazil is characterized by short spells of rain interspersed with frequent droughts. The hydrographic network of this region is seasonal and modest when compared to other Brazilian regions. The seasonality of the freshwater ecosystems is due to recurrence of extended drought, irregular rainfall, high temperatures and an elevated rate of water evaporation (Bouvy et al., 2000; Chellappa and Chellappa, 2004). This region receives almost 750 mm of annual rainfall and has an average annual temperature of 26°C throughout most of the region (Silva, 2004) and encompasses more than 750,000 km^2 in the northeast region of Brazil being the only Brazilian biome that is situated entirely inside the Brazilian territory (Leal et al., 2005). It is designated as a semiarid region due to its soil, vegetation, precipitation and temperature levels (Ab'Saber, 1995).

The abundance of water resources in Brazil does not imply that there is balanced water availability to the entire population. The semiarid freshwater ecosystems of northeast Brazil are represented by the rivers São Francisco and Parnaíba which are perennial, besides other few mid sized ones. These rivers have intermittent tributaries, and they flow to the Atlantic Ocean after crossing large dry areas of the semiarid land. Construction of reservoirs of varied sizes in Northeastern Brazil resulted in benefits, such as, potable drinking water facilities, development of inland fish culture, irrigation, small industries and agricultural activities. However, anthropic activities also resulted in environmental impacts, leading to loss of biodiversity, degradation of water quality and loss of native riverine fish species (Agostinho et al., 2005; 2008).

The freshwater ichthyofauna of the Neotropical region is characterized by high species richness which contributes to almost 31% of fish diversity of the planet (Reis et al., 2003). The fish communities that occur in the freshwater ecosystems of the tropical semiarid region of Brazil, exhibit a high degree of native species, which are the result of evolutionary processes mediated by climatic factors and hydrological cycles. However, there is need for more detailed information on diversity and distribution of fish fauna that occur in these ecosystems (Rosa et al. 2005; Chellappa et al., 2009a).

Studies have reported on different taxonomical groups of fish (Menescal et al. 2000; Rosa et al. 2005) and on their reproductive seasonality (Chellappa and Chellappa, 2004; Chellappa et al., 2009a) in the tropical semiarid Northeastern Brazil. These studies classified the semiarid fish species as native, invasive (species introduced or translocated beyond their natural ranges of occurrence within the Brazilian territory) and exotic (species introduced deliberately or accidentally from other countries).

Study Area and Sampling

Three major groups of freshwater ecosystems are encountered in tropical semiarid Brazil, the intermittent rivers and streams, lakes and reservoirs. These are disturbance-dominated systems with seasonal rains and frequent droughts being drivers of important processes maintaining diversity (Maltchik and Florin, 2002; Medeiros et al., 2010). The State of Rio Grande do Norte of the semiarid Northeastern Brazil, represents 52,797 km^2 (3.38%) of the Northeast area with 3 million inhabitants. Freshwater ecosystems in this state constitutes of seven drainage basins, the largest of which is the Piranhas-Assu hydrographic basin (latitudes 4° and 8° S and longitudes 36° and 39° W), encompassing 43,000Km^2. This river originates in the Serra do Bongo, southeast of the State of Paraiba, and enters the State of Rio Grande Norte, and eventually flows into the Atlantic Ocean. Several medium to small sized reservoirs have been constructed on this hydrographic basin, which are collectively responsible for a water storage capacity of 10 million m^3.

Rivers and reservoirs were selected in the Piranhas-Assu hydrographic basin to reflect anthropogenic disturbance (particularly impacts associated with fish cage culture practices, catchment land use associated with riparian and habitat degradation). Sites were sampled on a monthly basis during the period of 2008 to 2009, and fish were sampled in the reservoirs and downstream areas of the river that were wadeable. Samplings of 24 hours per day were conducted each month and fish samples were captured using fishing gear which consisted of stationary nets and gillnets of different mesh sizes (9 – 19 cm), cast nets, hooks and traditional traps.

Morphometric measurements and meristic counts were carried out to check the taxonomical status of each fish species. All fish collected were identified to the species level, measured (total body length to the nearest millimeter ±1 mm) and weighed (body mass±1 g). The constancy of fish species was calculated using the equation of Dajoz Index (Dajoz, 1973). $C = n / N \times 100$, where: C = constant; n = number of times the species was captured; N = total number of collections. Constant species (C> 50%), accessory species (25 <C <50%) and rare species (C <25) were calculated from the frequency of occurrence of each species in the sample. The results were compared with earlier studies (Menescal et al. 2000; Rosa et al. 2005).

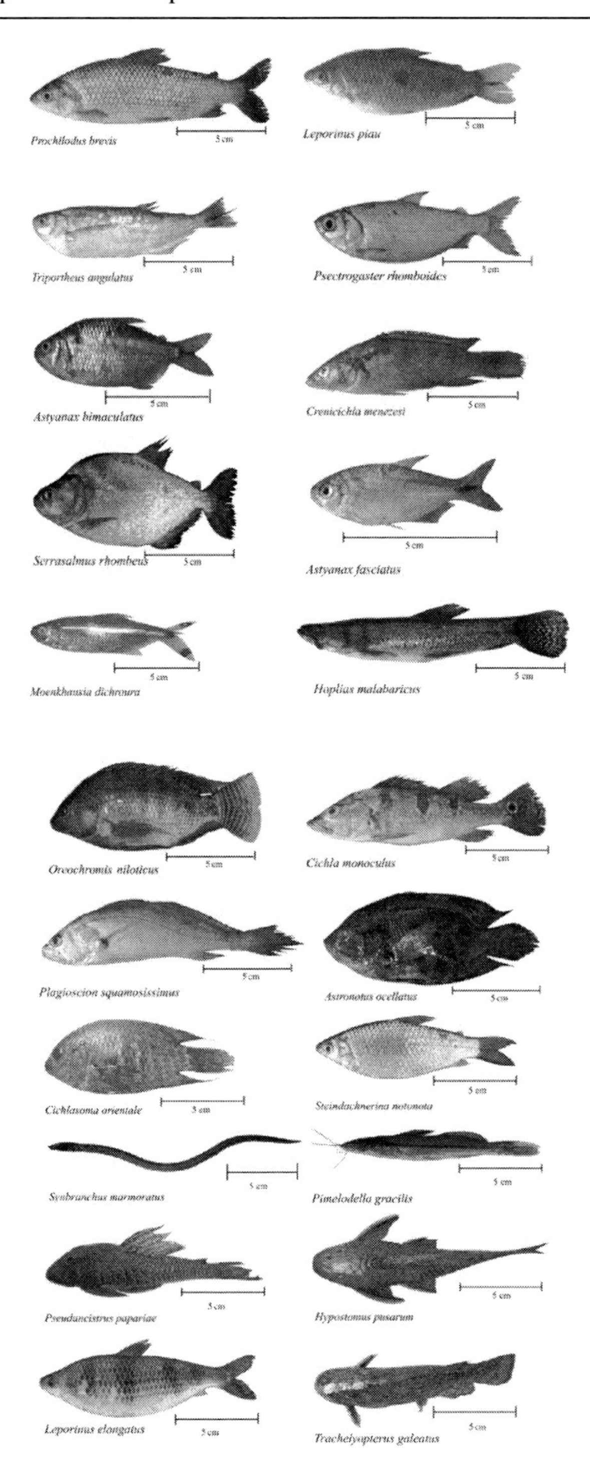

Figure 1. Fish species from the Piranhas-Assu hydrographic basin, Northeastern Brazil.

The mean monthly rainfall data of the study area, for the period 2000 to 2009, was obtained from the Meteorological Department of EMPARN (Empresa de Pesquisa Agropecuária do Rio Grande do Norte), Brazil.

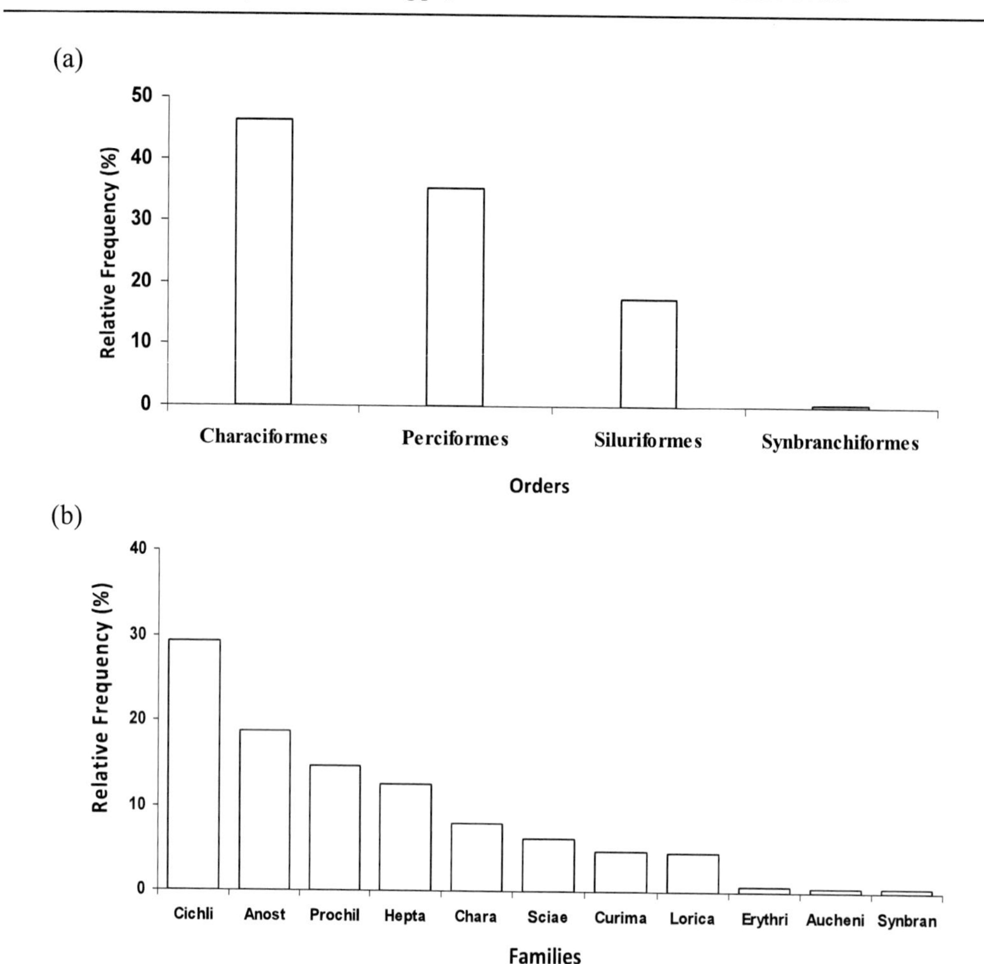

Figure 2.Relative frequency of (a) orders and (b) families of fish species from the Piranhas-Assu hydrographic basin of the semiarid Northeastern Brazil.

Results and Discussion

All fish species sampled belong to Osteichthyes (N=2604), and quantitative sampling of the fish fauna resulted in the collection of 22 species (Figure 1).

The twenty two fish species pertain to 4 orders (Characiformes, Perciformes, Siluriformes and Synbranchiformes) and 11 families (Characidae, Curimatidae, Auchenipteridae, Anostomidae, Prochilodontidae, Erythrinidae, Cichlidae, Sciaenidae, Heptapteridae, Loricariidae, Synbranchidae) (Figure 2).

Among the 22 fish species, 18 are native to semi arid region, three had been introduced from other hydrographic basins of Brazil and one was exotic. The representatives of the three following orders are all native fish species: Characiformes: *Astyanax bimaculatus* (Linnaeus, 1758), *Astyanax fasciatus* (Cuvier, 1819), *Moenkhausia dichroura* (Kner, 1858), *Triportheus angulatus* (Spix and Agassiz, 1829), *Serrasalmus rhombeus* (Linnaeus, 1766), *Psectrogaster rhomboides* (Eigenmann and Eigenmann, 1889), *Steindachnerina notonota* (Ribeiro, 1937), *Leporinus piau* Fowler, 1941, *Leporinus elongatus Valenciennes, 1850*, *Prochilodus brevis*

Steindachner, 1874, *Hoplias malabaricus* (Bloch, 1794); Siluriformes: *Trachelyopterus galeatus* (Linnaeus, 1766), *Pimelodella gracillis* (Valenciennes, 1835), *Pseudancistrus papariae* (Fowler, 1941), *Hypostomus pusarum* (Starks, 1913) and Synbranchiformes: *Synbranchus marmoratus* (Bloch, 1795).

The order Perciformes exhibited a mixture of endemic, introduced and exotic fish species. The native specieswere represented by *Cichlasoma orientale* Kullander, 1983 and *Crenicichla menezesi* Ploeg, 1991. The introduced fish species were represented by *Astronotusocellatus* (Agassiz, 1831), *Cichla monoculus* Spix and Agassiz, 1831, *Plagioscion squamosissimus* (Heckel, 1840) and the only exotic species was represented by *Oreochromisniloticus* (Linnaeus, 1758).

The orders which were more representative in number of fish species were as follows: Characiformes (46.35%), Perciformes (35.38%), Siluriformes (17.44%) and Synbranchiformes (0.50%)(Figure 2a). Among the families, Cichlidae was significantly representative (29.24%), followed by Anastomidae (18.77%)and the other nine families (Figure 2b). Among the fish species, the exotic fish Nile tilapia, *O. niloticus* was very expressive (24.92%) in the reservoirs, followed by the most common native species*L. piau* (18.44 %). These were followed by the other native and introduced species.

Table 1. Total length (Lt) in cm maximum, minimum and mean (±SD), weight (Wt) in gms maximum, minimum and mean (±SD) and percentage of occurrence (Ci) of fish species captured in Piranhas-Assu hydrographic basin, Rio Grande do Norte, Brazil

Species	Lt Min	Lt Max	Lt Mean ±SD	Wt Min	Wt Max	Wt Mean ±SD	Ci (%)
Astyanax bimaculatus	6.6	7.5	7.1 ±0.4	5.2	7.0	5.6±0.6	Constant (100)
Astyanax fasciatus	7.3	7.6	7.5±0.13	5.9	6	5.9±0.12	Accessory (40)
Moenkhausia dichroura	7.5	7.5	-	4	4	-	Rare (20)
Triportheus signatus	12	18	14.5±0.8	24	45.5	27.5±4.9	Constant (100)
Serrasalmus rhombeus	23	23	-	193.5	193.5	-	Rare (20)
Psectrogaster rhomboides	13.5	15.5	14.73±8.5	30	59.5	42.0±7.9	Constant (100)
Steindachnerina notonota	12	13	12.51±0.5	14	18	15.7±2.1	Accessory(40)
Leporinus piau	9,6	30	14,92±4,8	11,5	428	52,9±30,7	Constant (100)
Leporinus elongatus	33	33	-	540	540	-	Rare(20)
Prochilodus brevis	18	24.5	17.6±8.1	83	210	127.7±115	Constant (100)
Hoplias malabaricus	30	45	39.7±4.7	420	1000	825±170.8	Accessory (40)
Astronotusocellatus	22.5	22.5	-	248.5	248.5	-	Rare(20)
Oreochromisniloticus	19	30	17.5±7.1	132	520	162±14.5	Constant (100)
Cichlasoma orientale	12.3	15	13.7±1.2	44	78	61.1±10	Constant (60)
Crenicichla menezesi	9.5	18.5	13.94±1.9	8.5	69.5	32.7±13.2	Constant (80)
Cichla monoculus	13.5	27.5	21.1±5.5	65	310	166.6±110.2	Constant (80)
Plagioscion squamosissimus	14.5	34.5	24.5±5.5	74	525	195.3±135	Constant (100)
Trachelyopterus galeatus	15	20	17.7±2.5	68	125	95.8±28.3	Accessory (40)
Pimelodella gracillis	8.5	16.5	13±1.8	2.5	26	14.1±5.8	Constant (100)
Pseudancistrus papariae	18	21.5	19.5±2.1	95	205	130.9±53.5	Constant (60)
Hypostomus pusarum	17.5	22.5	9.1±2.4	91	203	129.7±52	Constant (60)
Synbranchus marmoratus	26.5	42	35.5±8	19	230	146.7±11.5	Accessory (60)

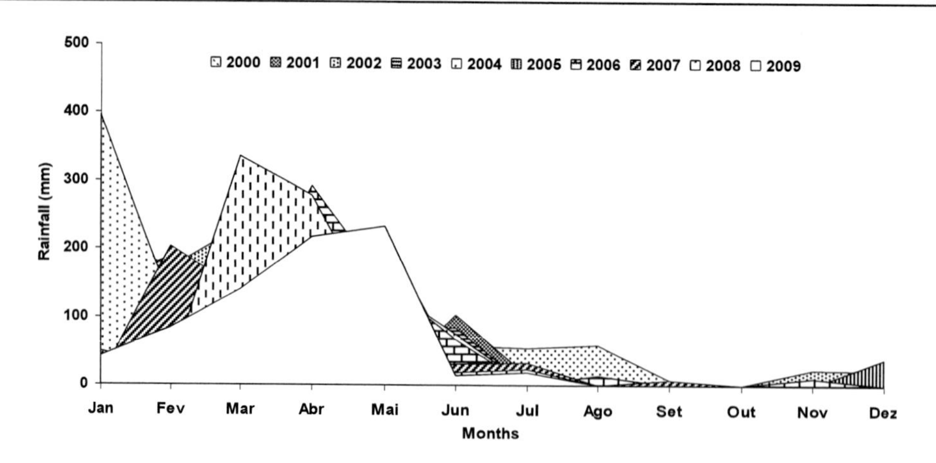

Figure 3.Mean monthly rainfall of the study area showing shifts in precipitation for the period 2000 to 2009.

Considering the relative frequency of occurrence of the 22 species, 13 were constant, five species were accessory and four were rare (*Moenkhausia dichroura*, *Serrasalmus rhombeus*, *Leporinus elongatus* and *Astronotusocellatus*). Table 1 shows the constancy of occurrence of fish species, maximum, minimum and mean (±SD) values of their total body length (Lt) and body mass (Wt).

Figure 3 shows the mean monthly rainfall of the study area and shifts in precipitation for the period 2000 to 2009. There is a shift in the period of rainy season in the semiarid region over the last ten years.

In the tropics as elsewhere, development is bringing changes that are potentially destructive to freshwater ecosystems. The increase in anthropic impacts over aquatic ecosystems in recent decades is an aggravating circumstance that has demanded the attention. Forests are being cleared, rivers are impounded and new agricultural practices are being introduced, including the massive use of agrochemicals. The sampling sites where introduced fish occur were characterized by significantly higher intensities of disturbance caused by human activities.

Earlier taxonomical surveys of fish fauna conducted in all the hydrographic basins of the entire Caatinga ecoregion registered the presence of 234 fish species, distributed in 8 orders (Rosa et al., 2005). The order Siluriformes had the highest diversity with 97 species followed by Characiformes (82), Cyprinodontiformes (28), Gymnotiformes (6), Perciformes (18), Myliobatiformes (1), Clupeiformes (1) and Symbranchiformes (1). Among the total fish species registered 9 were introduced and 136 were native to semiarid regions of Northeastern Brazil. In this chapter specific focus is given to the Piranhas-Assu hydrographic basin, wherein the order Characiformes has the highest fish diversity (11) followed by Perciformes (6), Siluriformes (4) and Synbranchiformes (1).

In South American aquatic systems the fish fauna is dominated by characoids (Characiforms) and siluroid catfishes (Siluriforms) and these two groups make up about 80% of the Amazonian fishes. These Neotropical fish represent one of the most extreme cases of evolutionary radiation and adaptation amongst living vertebrates (Goulding, 1980; Lowe-McConnell, 1987). The high frequency of occurrence of characoids and siluroids in the semiarid freshwater systems is in accordance with the ichthyofaunal distribution pattern encountered in the freshwater ecosystems of Brazil.

Menescal et al. (2000) examined the freshwater fish landings during the period of 1971 to 1998, which constituted of commercially important fish species from a reservoir in semiarid region. The native species were represented by *A.ocellatus, C. monoculus, P. brevis*(= *P. cearensis*), *P. squamosissimus, L. friderici, H. malabaricus* and exotic *O.niloticus*. However, *L. friderici* was not registered currently in the semiarid aquatic ecosystems of the State of Rio Grande do Norte.

Impacts Caused by Construction of Reservoirs on Small Rivers

In Brazil, the use of hydroelectric dams is considered the most viable way to produce electrical energy, besides the reservoirs are used for several purposes such as water storage for public usage, fish farming, tourism and leisure, contributing also to regional development (Tundisi and Matsumara-Tundisi, 2003). Modification of reservoir discharge alters the natural hydrological regime of water bodies located downstream. It reduces the mean annual discharge rate and the seasonal discharge rate, alters the time of occurrence of extreme discharge volumes, reduces the magnitude of floods and inflicts non-natural outflows. These changes interfere with abiotic factors, such as, water speed, substrate type, temperature and concentration of dissolved oxygen, which are important for aquatic organisms. The consequences of these changes include variation in water quality, decrease in dilution and natural capacity for purification, exposure of the river bed, and interference with input of allochthonous material originating from riparian vegetation (Poff et al. 1997). Thus, these changes in reservoir discharge rate modify the structure of fish communities and interfere with spawning of migratory species.

More than 15% of the Neotropical fish fauna is composed of migratory species (Carolsfeld et al., 2004), and the common reproductive strategy of the migratory fish involves upstream movements toward the spawning grounds at the beginning of the rainy season (Goulding, 1980; Chellappa et al., 1996; Agostinho et al., 2008). Reservoir constructions on rivers impact the migratory fish species pertaining to the orders of Characiformes and Siluriformes. However, reservoirs are characteristic features of the semiarid Northeastern Brazil located within the drought polygon. The impoundments on the hydrographic basins pose a major problem for the migratory fish populations. The characoid *P. brevis* is a rheophilic fish and depends on the extension of the river to complete spawning. Tragically in the semiarid regions there are no provisions for fish passages enabling this migratory species to complete its life cycle (Chellappa et al., 2009a). A series of reservoir constructions on a river can exclude migratory fish species from the basin or even lead to their complete extinction. Species of *Prochilodus* are an important ecological component of South American rivers. Taylor et al. (2006) investigated the effects of removing a dominant migratory detritivorous species, *P. mariae* on the functioning of the Las Marias ecosystem in the OrinocoBasin. The absence of this species caused changes in the metabolism and organic carbon flow of this ecosystem, leading to total degradation of the river.

IMPACTS CAUSED BY DEFORESTATION

Tropical forests are being degraded over very large spatial scales. Conservation agencies are interested in exploring deforestation impacts and most efforts have been directed towards understanding these impacts to terrestrial habitats and freshwater ecosystems. However, in the tropical semiarid region studies on impacts of deforestation on aquatic ecosystems are rather limited (Lacerda and Barbosa, 2006).

Land clearing for subsistence agriculture, fruit and plant cultivation, fire accidents induced by human activity and logging operations are principal activities which lead to the degradation of the aquatic systems and riparian vegetation. During the rainy season, the uprooted trees expose the soil to rains causing silting and consequently sediment runs off to the near by aquatic ecosystems.

Deforestation leads to changes in hydrodynamical regimes which cause negative impacts on fish communities (Welcomme, 1985). Removal of vegetation decreases the evapo-transpiration rate and the interception of precipitation. Sedimentation increases in the rivers following deforestation causing increased turbidity (Eckholm, 1976). Degradation is clearly demonstrated by the decrease in inland fishery production, and the reduction of the quality of water for public usage, irrigation systems and recreation.

IMPACTS CAUSED BY FISH CULTURE ACTIVITIES

The rapid increase in population in tropical semiarid regions, and the continued demand for fish, generated the need for inland fishculture as an alternative means of increasing fish production. Inland fishery activities are practiced in reservoirs and lakes (Chellappa et al., 1995). In aquatic systems the most significant effect of high rates of nitrogen input is eutrophication, which has a negative impact on fisheries and results in economic losses associated with degradation of water quality.

Local fisherman make indiscriminate use of fish feed for maintaining fish in cage culture practiced in reservoirs and lakes. Some improperly managed semiarid reservoirs tend to switch over from oligotrophic to eutrophic state through excess nutrient accumulation, emanated from unconsumed fish feeds.

The second dimension of negative impact is through the overstocking of the planktivorous fish, which stimulate the growth of unpalatable algal species. Such anthropic impacts result in the low water quality, diminished transparency and anoxic hypolimnion (Chellappa et al., 2009b).

The dominance of cyanobacterial species, particularly *Cylindrospermopsis raciborski* and *Microcystis aeruginosa*, have been registered in eutrophicated freshwater systems in the State of Rio Grande do Norte, in the form of sporadic events related to fish kills. The lethally affected fishes were *O. niloticus, P. squamosissimus, C. monoculus, P. brevis, H. malabaricus* and *Leporinus* spp. (Bouvy et al., 2000; Chellappa et al. 2000; Chellappa et al. 2008).

IMPACTS CAUSED BY INTRODUCTION OF NATIVE AND EXOTIC FISH

The Amazonian fish of commercial importance were introduced into the reservoirs of the semiarid northeastern regions of Brazil, since they exhibit a marked degree of physiological plasticity that allows them to adapt to a highly variable natural environment (Val and Almeida-Val, 1995; Val et al., 1996; Chellappa et al., 2003). The introduced fish species were *Cichla monoculus*, *Astronotus ocellatus*, *Colossoma macropomum*, *Piaractus brachypomum*, *Arapaima gigas* and*Plagioscion squamosissimus*, besides the exotic species *Tilapia rendalli* and *Oreochromis niloticus*.Among the introduced fish species *O. niloticus, C. monoculus* and *P. squamosissimus* are well established (Chellappa, 2000). Transplanting species from one basin to the other can represent a menace to the local species. Introduced fish may be better adapted to or more tolerant of the environmental conditions, thus achieving higher growth and reproductive output, as in the case of *C. monoculus* (Chellappa et al., 2003).

Exotic Nile tilapia, *O. niloticus* has been introduced in the freshwater ecosystems of the Caatinga region since the beginning of 1970 in order to upgrade the fish production and fish culture (Gurgel and Fernando, 1994). Presently, it is a dominant species in most reservoirs of northeastern Brazil. This is due to its rapid growth, omnivorous food habits, high reproductive efficiency, parental care and resistance to environmental variations. The total gross production of tilapia increased exponentially from 1970 to 2000.Menescal et al. (2000) observed changes in fish community structure after the introduction of Nile tilapia in semiarid Rio Grande do Norte, Brazil, with damages caused to the native species like,*P. brevis, Leporinus* spp. and *H. malabaricus*. The proliferation of exotic species in the freshwater ecosystems of semiarid region is considered a threat to the integrity of native fish community. Tragically, fish culture practices in the Brazilian semiarid region has followed suite to meet consumer aspirations through a moderate shift into the culture of exotic Nile tilapia.

A new predator in freshwater ecosystems often results in a total change in community structure, as it had occurred with the introduction of the Nile perch, *Lates niloticus* in Lake Victoria of East Africa; the introduction *Cichla monoculus* in Lake Gatun in Panamaand a goby, *Glossogobius giurus*, in Lake Lanao (Payne, 1986).

One of the most important consequences of exotic and predatory fish species introductions in the semiarid region has led to partial or total disappearance of native fish species. This could be followed by alterations in the trophic chain, balance of natural populations, or ecological processes. Native fish species show reduction in diversity between 2000 and 2009, with the disappearance of a native species. Furthermore, the remaining species show reduced abundance.These changes are attributed to the introduction of an exotic competitor species like Nile tilapia, *O. niloticus* and predator species like *C. monoculus* and *P. squamosissimus*.

IMPACTS CAUSED BY SHIFTS IN RAINFALL

Life cycle characteristics such as reproductive seasonality of fish species are regulated by environmental drivers. They influence gonadal development phases, synchronize processes involved in final gamete maturation, signal ideal conditions for spawning and reproductive activity. Reduced fluctuation in water temperature and photoperiod may cause fish species in tropical environments to utilize alternate environmental drivers to initiate reproductive activities during suitable periods of the year. Northeast Brazil represents the semiarid tropical belt where changes in rainfall regimes cause some seasonality. Rainfall is considered as the most important driver for reproductive activity of fish, which modulates the timing of reproduction (Lowe-McConnell, 1987; Chellappa et al., 2003). Spawning seasons of semi-arid freshwater fish fauna are either restricted to rainy season as in the potamodromous *P. brevis*, or extended as in the cichlids, with spawning occurring during most part the year. These species respond in an individualistic manner to changes in rainfall with consequent changes in their reproductive plasticity(Chellappa et al., 2009a).

Indiscriminate fishing occurs when the migratory fish species reproduce under the influence of rainfall. There is a need to reconsider the out of season fishing period which are currently enforced during fixed times. The spawning period of fish vary from year to year coinciding with the shifts in rainfall rhythms, which should be taken into consideration by administrative agencies while enforcing the out of season fishing period.

Conclusion

The conservation of fish diversity is a major global environmental challenge. Serious and urgent administrative action is needed to curb anthropogenic activities, such as non sustainable aquaculture practices; release of agro toxins, urban and industrial sewage effluents in aquatic ecosystems; the construction of many reservoirs and dams on hydrographic basins; destruction of riparian vegetation; introduction of exotic and predatory fish species and indiscriminate fishing during the reproductive season of migratory fish. These factors are the main causes of loss of fish species.Strict regulations on introductions of fish species and restrictions on release of pollutants to aquatic ecosystems would reverse ecosystem degradation, improve water quality and conserve fish diversity.

Acknowledgment

This study was funded by the National Council for Scientific and Technological Development of Brazil (CNPq). Our thanks are due to Dr. Ricardo Souza Rosa of the Universidade Federal da Paraíba, Brazil, for the collaboration given in the taxonomical identification of fish.

References

Ab´Saber, A.N. The Caatinga Domain. In: Monteiro, S. and Kaz, L., editors, Caatinga-Sertão, Sertanejos. Rio de Janeiro: *Editora Livroarte;* 1995; 47-55.

Agostinho, A.A.; Pelicice, F.M. and Gomes, L.C. (2008). Dams and the fish fauna of the Neotropical region: impacts and management related to diversity and fisheries. *Brazilian Journal of Biology*, 68, 1119-1132.

Agostinho, A.A.; Thomaz, S.M. and Gomes, L.C. (2005). Conservation of the biodiversity of Brazil's inland waters. Conservation Biology,19, 646-652.

Andrade-Lima, D. (1981). The caatingas dominium. *Revista Brasileira de Botânica*, 4, 149-163.

Bouvy, M.; Falção, D.; Marinho, M.; Pagano, M. and Moura, A. (2000). Occurrence of *Cylindrospermopsis* (Cyanobacteria) in 39 Brazilian tropical reservoirs during the 1998 drought. *Aquatic Microbial Ecology,*23, 13-27.

Carolsfeld, J.; Harvey, B.; Ross, C.; Baer, A. (2004). Migratory fishes of South America: biology, fisheries and conservation status. *Ottawa:* World Fisheries Trust, World Bank, IDRC.

Chellappa, S. (2000). A Review on reproductive strategies and ecology of cichlid fishes in Northeastern Brazil. *Journal of Tropical Aquatic Ecology,* 10, 5-11.

Chellappa, S. andChellappa, N.T.Ecology and reproductive plasticity of the Amazonian cichlid fishes introduced to the freshwater ecosystems of the semi-arid Northeastern Brazil. In: Kaul,B.L., editor. *Advances in Fish and Wildlife Ecology and Biology* (Vol. 3). New Delhi: Daya Publishing House; 2004; 49 – 57.

Chellappa, S.; Câmara, M.R. and Chellappa, N.T. (2003). Ecology of *Cichla monoculus* (Osteichthyes: Cichlidae) from a reservoir in the semiarid region of Brazil. *Hydrobiologia,* 504, 267-273.

Chellappa, N. T.; Chellappa, S. L. and Chellappa, S. (2008). Harmful phytoplankton blooms and fish mortality in a eutrophicated reservoir of Northeastern Brazil. *Brazilian Archives of Biology and Technology,* 51 (4), 833 - 841.

Chellappa, N.T.; Costa, M.A.M. and Marinho, I.R. (2000). Harmful cyanobacterial blooms from semiarid freshwater ecosystems of Northeast Brazil. *Australian Society Limnology Newsletter,* 38, 45-49.

Chellappa, S.; Cacho, M. S. R. F.; Huntingford, F. A.and Beveridge, M. C. M. (1996). Observations on induced breeding of the Amazonian fish tambaqui, *Colossoma macropomum* (Cuvier) using CPE and HCG treatments. *Aquaculture Research,* 27 (2), 91-94.

Chellappa, S.; Chellappa, N. T.; Barbosa, W. G.; Huntingford, F. A.and Beveridge, M. C. M. (1995). Growth and production of the Amazonian tambaqui in fixed cages under different feeding regimes. *Aquaculture International,* 3 (1), 11-21.

Chellappa, S.; Bueno, R.M.X.; Chellappa, T.; Chellappa, N.T. and Val, V.M.F.A. (2009a). Reproductive seasonality of the fish fauna and limnoecology of semi-arid Brazilian reservoirs. *Limnologica*, 39, 325- 329.

Chellappa, N.T.; Chellappa, T.; Câmara, F.R.A.; Rocha,O. and Chellappa, S. (2009b). Impact of stress and disturbance factors on the reservoir phytoplankton communities in Northeastern Brazil. *Limnologica,* 39, 273-282.

Eckholm, E.P. (1976). *Losing Ground: Environmental stress and world food prospects.* New York: Norton.

Goulding, M. (1980).*The fishes and the forest: explorations in Amazonian natural history.* Los Angeles: University of Califórnia Press.

Gurgel, J.J.S. and Fernando, C.H. (1994). Fisheries in semiarid northeast Brazil with special reference to the role of tilapias. *Hydrobiologia,* 79, 77–94.

Lacerda, A.V. and Barbosa, F.M. (2006). *Matas ciliares no domínio das caatingas.* João Pessoa: Editora Universitária, Universidade Federal da Paraíba.

Leal, I.R.; Silva, J.M.C.; Tabarelli, M. and Lacher Jr., T.E. (2005). Changing the course of biodiversity conservation in the Caatinga of Northeastern Brazil. *Conservation Biology,* 19, 701- 706.

Leal, I.R.; Tabarelli, M. and Silva, J.M.C. (2003). *Ecologia e conservação da Caatinga.* Recife: Editora Universitária, Universidade Federal de Pernambuco.

Leite, A.V.L. and Machado, I.C. 2010. Reproductive biology of woody species in Caatinga, a dry forest of northeastern Brazil. *Journal of Arid Environments,* 74, 1374 -1380.

Lockwood, J.L., Hoopes, M.F. and Marchetti, M.P. (2007). *Invasion Ecology.*Oxford: Blackwell Publishing.

Lowe-McConnell, R.H., (1987). *Ecological Studies in Tropical Fish Communities.* Cambridge: CambridgeUniversity Press.

Maltchik, L. and Florin, M. (2002). Perspectives of hydrological disturbance as the driving force of Brazilian semiarid stream ecosystems. *Acta Limnologica Brasiliensia,* 14 (3), 35 - 41.

Matthews, W.J. (1998). *Patterns in Freshwater Fish Ecology.*New York: Chapman and Hall.

Medeiros, E.S.F.; Silva, M.J.; Figueiredo, B.R.S.; Ramos, T.P.A. and Ramos, R.T.C. (2010). Effects of fishing technique on assessing species composition in aquatic systems in semiarid Brazil. *Brazilian Journal of Biology,* 70 (2), 255 – 262.

Menescal, R.A.; Oliveira, J.C.S.; Campos, C.E.C.; Araújo, A.S. and Freire, A.G. (2000). Fish production in Marechal Dutra Reservoir, Acari, RN. *Journal of Tropical Aquatic Ecology,* 10, 135-139.

Nelson, J.S. (1994). *Fishes of the World.*New York: John Wiley and Sons.

Poff, N. L.;Allan, D.;Bain, M. B.; Karr, J. R.;Prestegaard, K. L.; Richter, B.D.;Sparks R.E. and Stromberg. C. (1997). The natural flow regime: a paradigm for river conservation and restoration. *Bioscience, 47* (11), 769-784.

Payne, A.I. (1986). *The ecology of tropical lakes and rivers.* New York: John Wiley and Sons.

Queiroz, L. P., Rapini, A. and Giulietti, A. M. (2006). *Towards greater knowledge of the Brazilian Semiarid Biodiversity.* Brasília: Ministério da Ciência e Tecnologia.

Reis, R. E., Kullander, S. O. and Ferrari Jr. C. J. (2003). *Check list of the freshwater fishes of South and Central America.* Porto Alegre: EDPUCRS.

Rosa, R.S.; Menezes, N.A.; Britski, H.A.; Costa, W.J.E.M. and Groth, F. Diversidade, padrões de distribuição e conservação dos peixes da Caatinga. In: Leal, I.R.; Tabarelli, M. and Silva, J.M.C., editors. *Ecologia e Conservação da Caatinga. Recife: Editora UFPE;* 2005; 135-180.

Silva, V.P.R. (2004). On climate variability in Northeast of Brazil. *Journal of Arid Environments,* 58, 575 - 596.

Taylor, B.W.; Flecker, A.S. and Hall Jr., R.O. (2006). Loss of a harvested fish species disrupts carbon flow in a diverse tropical river. *Science,* 313, 833-836.

Tundisi J.G. and Matsamura-Tundisi T. (2003). Integration of research management in optimizing multiple uses of reservoirs: the experience in South America and Brazilian case studies. *Hydrobiologia,* 500, 231-242.

Val, A.L. and Almeida-Val, V.M.F. (1995). *Fishes of the Amazon and their environment: Physiological and Biochemical aspects.* Heidelberg: Springer Verlag.

Val, A.L., Almeida-Val, V.M.F. and Randall, D.J. *Physiology and biochemistry of the fishes of the Amazon.*In: Val, A.L.; Almeida-Val, V.M.F. and Randall D. J., editors. Physiology and Biochemistry of the Fishes of Amazon. *Manaus:* INPA; 1996; 1-3.
Welcomme, R.L. (1985). *River fisheries.*Rome: FAO Fisheries Technical Paper 262, 330p.

Reviewed by Professor Dr. José Zanon de Oliveira Passavante. Department of Oceanography, Universidade Federal de Pernambuco, UFPE, Recife, CEP: 50670-901, PE, Brazil.

In: Fish Ecology
Editor: Sean P. Dempsey

ISBN 978-1-61324-282-7

Chapter 7

FISH ECOLOGY, CONSERVATION BIOLOGY, AND NEW INSIGHTS FROM THE ARCHAEOLOGICAL EVIDENCE IN THE BEAGLE CHANNEL (TIERRA DEL FUEGO, ARGENTINA)

***Atilio Francisco Zangrando*[1] *and María Paz Martinoli*[2]**
[1]CADIC-CONICET / Universidad de Buenos Aires, Argentina
[2]Asociación de Investigaciones Antropológicas - Universidad de Buenos Aires, Argentina

ABSTRACT

This chapter analyses the taxonomic representations of marine fish from eight archaeological assemblages of the Beagle Channel.This study provides information about fish taxonomic distribution and ecological conditions in this region since 6400 to 500 radiocarbon years BP. Species corresponding to Nototheniidae family are represented throughout the archaeological sequence, in particular *Paranotothenia magellanica* and species attributable to *Patagonotothen* gender. Other fish of intertidal or shallow waters, as *Austrolycus depressiceps*, *Cottoperca gobio*, *Sprattus fueguensis*, etc., are also presented in the archaeological record. Among deep-sea species, *Macruronus magellanicus* is widely represented in the archaeofaunal assemblages. Nevertheless, the data also indicate that many species of deep waters (*Merluccius* sp. and *Thyrsites atun*)frequently consumed by hunter-gatherers are not common today in the marine ecosystem of the Beagle Channel. Comparisons of these data with modern ecological surveys indicate that both environmental stabilityat coastal waters and changes of the distributions of deep-sea species have occurred. This last result provides a new insight regarding the influence of modern human activities and the scope of overfishing in the uttermost part of the world.

INTRODUCTION

The use of zooarchaeological information of fish to provide a historical perspective to wildlife managers and conservation biologists has been significantly increased during the last decade (Jackson *et al.* 2001; Lyman and Cannon 2004;Reitz 2004; Jones 2007; Reitz*etal.* 2009, among others). Conservation biologists typically need a framework of ecological base for recreating or maintaining the natural landscape in an area. In this regard, this frame of reference is a goal of conservation activities; it is an ecological condition which is desired (Lyman 2006). Zooarchaeological record provides data about the long-term functioning of an ecosystem that combine many ecological and pre-industrial anthropogenic processes, and which can serve as a guide in establishing a frame of ecological reference.

Archaeological sites of the Beagle Channel (Tierra del Fuego, Argentina) contain a record of changes and stability of fish ecology and fish use by humans to Middle and Late Holocene, which could be a great advantage to the conservation effort by establishing a historical base for fishing modern management decisions. Nevertheless, interpretations based on zoooarchaeological evidence attending to this important issue are not simple and require some cautions, which are well known by the archaeologists, but the particularities of this record need to be clarified for a different audience (Lyman 1996; Reitz 2004). Zooarchaeological assemblages cannot be used as mirrors of the distribution and/or abundance of animal resources in the past (Reitz 2004). These records are the result of a series of cultural and natural processes that act during and/or subsequently to the deposition of the faunal remains. The zooarchaeological record reflects choices made by human groups with dietary, technological or ideological goals, and some resources available in the environment could be used while others might be ignored. Exploitation of resources could also have responded to certain cycles in the availability of resources, usually measured in seasonal or annual scales, or to technological capabilities for their harvest or processing. In turn, fish bones are exposed to taphonomic processes which affect the taxonomical composition and generally lead to bone loss in the archaeological record. All these aspects should be considered when we evaluate the archaeological evidence with a paleoecological perspective.

The main goal of this chapter is to assess long-term changes in availability of marine fish and ecological abundance in the Beagle Channel region. The diversity and availability of various fish species of this region is discussed in relation to location and temporal variability from a historical perspective. To achieve this, we use archaeological data to trace patterns in resource use by pre-industrial societies from 6400 to 500 radiocarbon years before present (BP), and we compare this information with modern biological records on the current taxonomic composition and distribution of fish in the Beagle Channel.

In this chapter we take the entire span of the archaeological sequence of the Beagle Channel as a single unit of analysis. We agree with Reitz *et al.* (2009) in that it is necessary to adopt a regional scale to distinguish anthropogenic and/or non-anthropogenic patterns of historical variations. Only in aregional scope we canbe morecertainthatthe observed patterns which do not answer tolocal or specific behavioral events of individuals or communities, or toparticulartaphonomichistories.

The Study Area

The Beagle Channel is located at the southern tip of the Americas (54° 50'S, between 66° 30' and70° W; Figure 1). With approximate 180 km of length, it runs east to west with a width ranging from four to seven kilometers. It is a glacial valley that has been invaded by sea water about 8000 years ago (Rabassa *et al.* 1986), thus conforming a channel of communication between Pacific and Atlantic oceans. Parallel to the north coast runs a chain of mountains (Fuegian Andes) with altitudes up to 2500 ma.s.l. The proximity of the mountain range becomes abrupt various sectors of the coast and the range affected by the tides is narrow, with the subsequent rapid deepening of the seabed (Figure 2).

The marine waters of the Beagle Channel have low salinity, with a seasonal variation between July (31.3 ‰) and November-December (26.5 ‰). The average annual temperature of seawater is 6.3 °C, varying between 8.6 °C in January and 4.3 °C in August (Iturraspe and Schroeder, in Orquera and Piana 1999a). The waters do not freeze in winter, except in some very small inlets or bays. Prevailing winds blow from southwest or west, which coincide with the geographical orientation of the channel.

The annual average temperature (air) was calculated for the entire 20th century by 5.3 °C, with an average range of variation ranging from 9.3 °C in January to 1.1 ° C in July (Iturraspe and Schroeder, in Orquera and Piana 1999a). The average annual rainfall between 1901 and 1996 determined is 530 mm. In the winter the snow component is high, despite the rain alternating with snow. Both shores of the Beagle Channel provide a large drainage network, which consists of rivers, creeks and small streams of water that are fed by rain, melting snow on high mountain and glacial contributions.

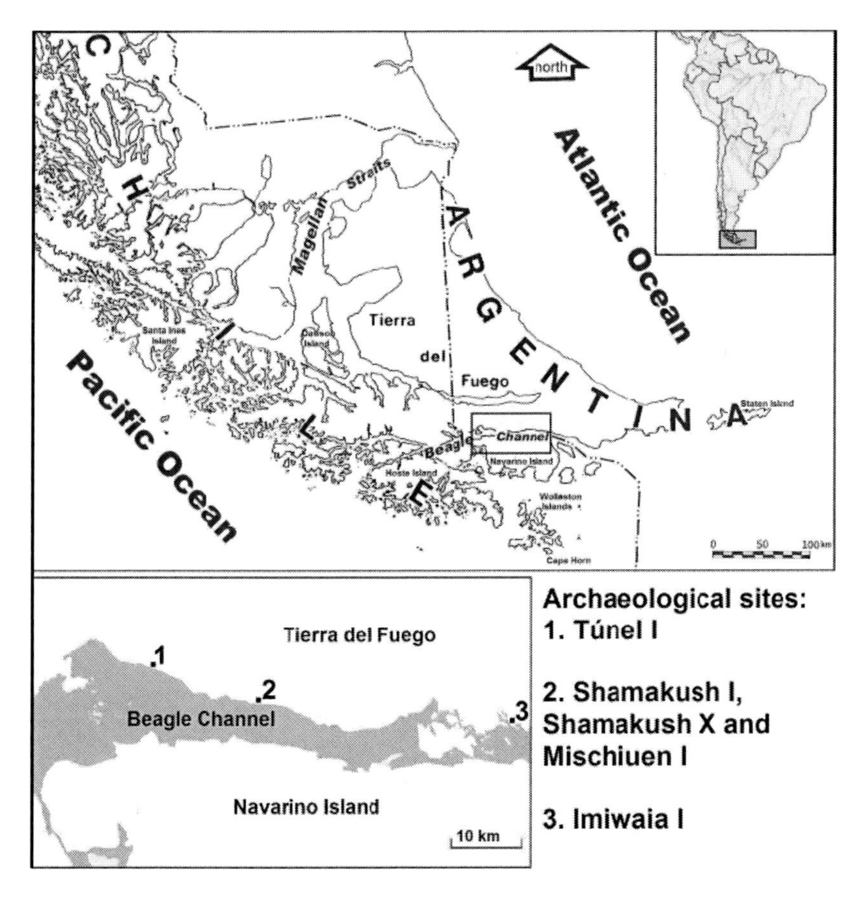

Figure 1. Beagle Channel and locations of the archaeological sites.

Figure 2. Panoramic view of the Beagle Channel.

Modern Records of Fish Fauna

There are three inventories on fish fauna of the Beagle Channel, which comprise a book (Lloris and Rucabado 1991) and two reports (Fenucci *et al.* 1978; Lopez *et al.* 1996).There is also information from other articles on aspects of distribution and ecology of the species present (Moreno and Jara 1984; Rae 1991; Vanella *et al.* 2007; among others). Based on these papers, then we develop a description of the taxonomic composition and distribution of fish fauna of the Beagle Channel.

The samplings reported in those studies have some differences in relation to locations, moments of year and equipment used in catches, which is important when we evaluate the presence, abundance and distribution of resources in the environment. Fenucci *et al.* (1974) carried out 15 sampling operations along the Beagle Channel in February and March 1974, but only on deep sea benthic trawling. Lloris and Rucabado(1991)reportedthe resultsof 20samples takenindifferentmonths in1987and indifferentmicroenvironmentsof the Beagle Channel. The captures were held by hooks (longline), nets (trammel nets, and beach nets) and by hand (using chemicals anesthetic –quinaldine-), plus two benthic trawling (Lloris and Rucabado 1991).In addition, these researchers report the results of samples taken in February and March of 1988, with similar technologies but with the exception of benthic trawling. López and coauthors (1996) conducted monthly sampling in 20 stations between February 1987 and March 1988. They used different types of methods to capture, among others, trawls, vertical and horizontal longlines, trammels and anesthetics.In all these investigations catch records were mainly expressed as number of individuals or as biomass (kg).

In the waters of the Beagle Channel, it was identified a total of 56 species, grouped in 23 families and 9 orders (Table 1). Moreno and Jara (1984) analyzed the distribution of species in shallow water in two types of seabed. On one hand, rocky shores make substrates that allow the consolidation of *Macrocystis pyrifera* kelp. These kelps may form extensive "forests" located below the lower limit of the intertidal zone to 30 m deep. On the other, sandy bottoms are more homogeneous substrates and do not permit the colonization of this kelp.

Table 1. Marine fish taxa present in the Beagle Channel

Class	Order	Family	Species
OSTEI-CHTHYES	CLUPEI-FORMES	Clupeidae	*Sprattus fueguesis*
	SALMO-NIFORMES	Galaxiidae	*Galaxia maculatus*
		Aplochitonidae	*Aplochiton taeniatus*
	GADI-FORMES	Moridae	*Salilota australis*
		Merluccidae	*Macruronus magellanicus*
			Merluccius hubbsi
			Merluccius australis
		Ophidiidae	*Genypterus blacodes*
		Zoarcidae	*Austrolycus depressiceps*
			Crossostomus chilensis
			Crossostomus sobrali
			Haushi marinae
			Ilucoetes facali
			Ilucoetes fimbriatus
			Maynea patagonia
			Phucocoetes latinans
		Macrouridae	*Macrourus holotrachys*
			Coelorrhynchus fasciatus
	ATHERI-NIFORMES	Atherinidae	*Odontesthes nigricans*
	SCORPAE-NIFORMES	Scorpaenidae	*Sebastes oculatus*
		Congiopodidae	*Congiopodus peruvianus*
		Agonidae	*Agonopsis chiloensis*
		Cyclopteridae	*Careproctus pallidus*
	PERCI-FORMES	Carangidae	*Parona signata*
		Bovichthyidae	*Cottoperca gobio*
		Notothenidae	*Dissostichus eleginoides*
			Eleginops maclovinus
			Harpagifer bispinnis
			Harpagifer georgianus georgianus
			Notothenia trigramma
			Paranotothenia angustata
			Paranotothenia magellanica
			Patagonotothen brevicauda
			Patagonotothen cornucola
			Patagonotothen longipes
			Patagonotothen ramsayi
			Patagonotothen sima
			Patagonotothen tessellata
			Patagonotothen canina
			Patagonotothen wiltoni
		Chaennichthydae	*Champsocephalus essox*
		Gobiidae	*Ophiogobius ophicephalus*
		Gempylidae	*Thyrsites atun*
		Stromateidae	*Stromateus brasiliensis*
CHONDRI-CHTHYES	SQUALI-FORMES	Squalidae	*Centroscyllium fabrici*
			Squalus acanthias
	RAJIFORMES	Rajidae	*Bathyraja albomaculada*
			Bathyraja brachyurops
			Bathyraja griseocauda
			Bathyraja magellanica
			Bathyraja scaphiops
			Psammobatis rudis
			Psammobatis scobina
			Raja (Dipturus) trachyderma
			Raja (Dipturus) flavirostris
	CARCHA-RINIFORMES	Scylliorhinidae	*Schroederichthys bivius*

Information was taken from Fenucci *et al.* 1978; Lloris and Rucabado 1991, Lopez *et al.* 1996.

From their observations, Moreno and Jara (1984) inferred that where the density of kelps (*M. pyrifera*) is larger, greater is the abundance of fishes in shallow waters (particularly *Paranotothenia magellanica* and different species of the genus*Patagonotothen*). Studies conducted in 1987 and 1988 by Lopez *et al.* (1996) are consistent with these observations. This is because the ecosystem of kelps has two conditions that positively influence the adaptive adjustment of the Nototheniid fish (Moreno and Jara 1984; Rae 1991; Vanella *et al.* 2007): 1. food availability (their diet is basically the consumption of amphipods and isopods associated with *M. pyrifera*), and 2. refuge of predators.

Eleginops maclovinus is perhaps the most representative species of the Beagle Channel by distribution and abundance (Lloris and Rucabado 1991, López *et al.* 1996; Figure 3). This species runs along the shoreline at shallow depth and usually enters in the rivers. *Sprattus fueguensis* is a pelagic species with some tolerance to low salinities. Large shoals of this species can beach in bays with some frequency (Lloris and Rucabado 1991). Species of the Zoarcidae family can be found frequently in the intertidal zone.

Macruronus magellanicus commonly inhabits depths between 30 and 500 meters, although youth individuals may frequent coastal waters (Leible *et al.* 1981). In the Beagle Channel, this species can be captured abundantly from the coast at depths less than 1 meter to 110 meters of deep, regardless of their size and maturity (Lloris and Rucabado 1991). Its presence in the Beagle Channel is recorded usually between January and April (Lopez *et al.* 1996). Although less common, two species of the genus*Merluccius* (*Merluccius australis* and *Merluccius hubbsi*) can also be captured in depths more than 30 meters. Other species inhabit deep waters all year round; among the most characteristic species are *Salilota australis*, *Genypterus blacodes*, *Dissotichus eleginoides* and several species of Chondrichthyes (Lloris and Rucabado 1991).

Zooarchaeological Samples and Methods

In this chapter we analyze fish bone remains recovered in eleven archaeological assemblages with chronologies which span from 6400 to 500 radiocarbon years BP. Radiocarbon ages for each of these assemblages are presented in Table 2. The archaeological record of the Beagle Channel is the result of lifestyle and activities developed by groups of hunter-gatherers (Orquera and Piana 1999b). The material culture recorded in archaeological assemblages and in ethnographic sources shows that these groups used essentially two types of technology for fishing: harpoons and fishing lines (Orquera and Piana 1999a, 1999b). This differs from the variability and extent of methods used in modern samplings, which should be considered in comparisons between the two types of records.

The recovery strategies of ichthyofaunistic remains analyzed in this chapter were very similar between the different archaeological sites. The methods used were the direct recovery and screening of sediments through sieves of 2 mm and 5 mm of aperture.

The taxonomic identification of fish remains was carried out using the method of comparative anatomy for total specimens recovered in archaeological sites. The identification was made mostly by macroscopic observations and in some cases using a stereomicroscope. To establish the representation of different taxa identified, two common measures in zooarchaeology were used (Grayson 1984; Reitz and Wing 1999): the Number of Identified Specimens (NISP) and the Minimal Number of Individuals (MNI).

The procedure adopted to evaluate bone fragmentation in ichthyo-archaeological assemblages was the proposed by Zohar *et al.* (2001). In this case, a degree of representation using a percentage scale of five intervals is assigned to each bone, which indicates the proportions of the represented bones: 1. complete (100 - 91%) 2. slightly fragmented (90 - 71%) 3. partially fragmented (70 - 51%), 4. highly fragmented (50 - 26%) and 5. fragment (25% or less). In order to standardize this information, then we calculated an average rate of fragmentation by the following formula (Zohar *et al.* 2001):

$$\sum (Wi * Xi)/100$$

where Wiis the proportion of bones recorded for each interval of fragmentation and Xi represents the fivecategories within each of the intervals (in this case 100%, 80%, 60%, 40% and25%).

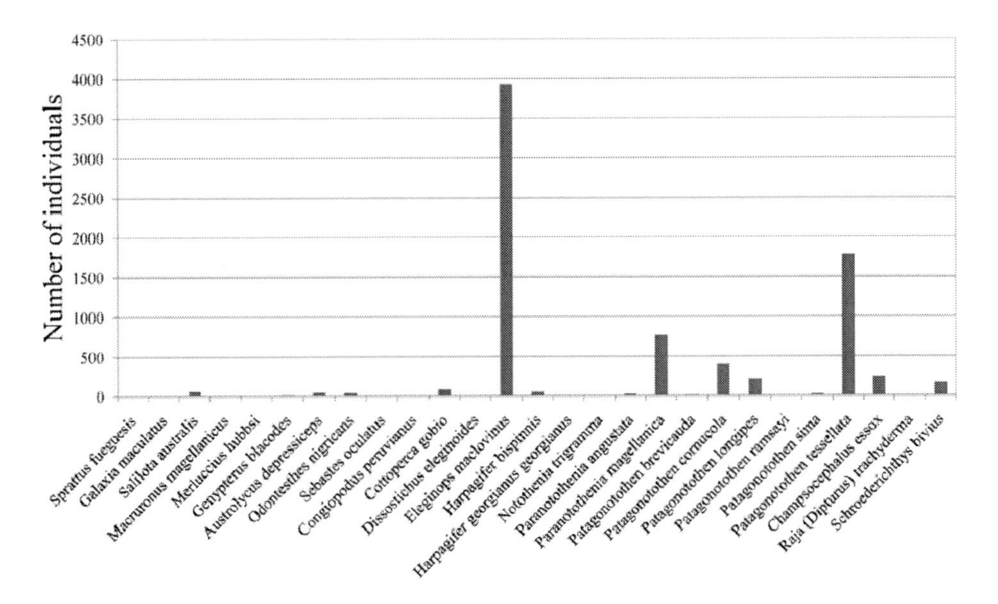

Figure 3. Relative abundances of marine fish in the Beagle Channel; information taken from Lopez *et al.* (1996).

The discussion of this chapter is based in part on the comparison between the currently diversity of fishes in the Beagle Channel and the taxonomical compositions of zooarchaeological assemblages. To know the degree of similarity, both records were taken to percentages and also applied the Jaccard index (Magurran 1989):

$$Cj = \frac{C}{(A+B) - C}$$

where A is the number of families or species currently present in the Beagle channel, B is the number of families or species present in the archaeological samples and C the number of species common to both records.

ZOOARCHAEOLOGICAL DATA

Table 3 shows the numbers of identified specimens (NISP) and minimum number of individuals (MNI) estimates for different taxa presented in the archaeological sequence of Beagle Channel. MNI values were standardized and presented in relative frequencies in figures 4 to 7; the assemblages are arranged chronologically beginning with the earliest next to the vertical axis. We identified 11 species from 9 families of bony fish; at the moment, cartilaginous fish were not identified at archaeological sites of the Beagle Channel.

Table 2.Archaeological assemblages analysed in this chapter

Site	Stratigraphic Unit	Age (^{14}C years BP)
Túnel I	Second Component	6470 ± 100 - 4590 ± 130
	Third Component	4300 ± 80 AP
	Fourth Component	2690 ± 80 - 2660 ± 100
	Fifth Component	1990 ± 110
	Sixth Component	670 ± 80 - 450 ± 60
Imiwaia I	Layers K, L and M	5872 ± 147 - 4900 ± 120
	Layer D	3013 ± 38
	Layer B	1577 ± 41
Mischiuen I	Layer F	4890 ± 210 - 4430 ± 130
Shamakush I		1220 ± 110 - 940 ± 110
Shamakush X		500 ± 100

Chronological information was taken from Orquera and Piana 1999a, and Zangrando 2009a.

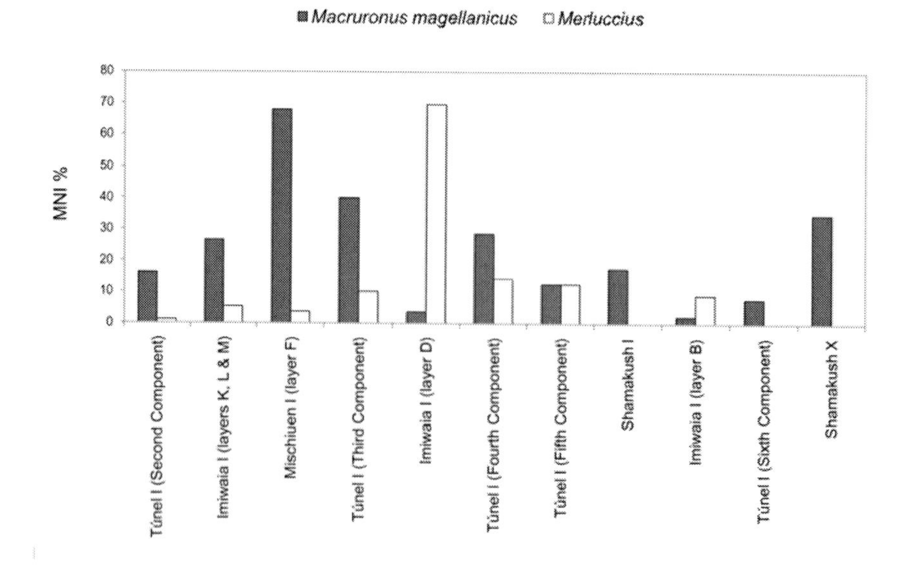

Figure 4. Relative abundances of *Macruronus magellanicus* and *Merluccius* throughout the archaeological sequence.

Table 3.Numbers of identified specimens and minimum numbers of individuals obtained inichthyoarchaeological assemblages of the Beagle Channel

Fish Taxa	Túnel I (Second Component)		Imiwaia I (layers K,L and M)		Mischiuen I (layer F)		Túnel I (Third Component)		Imiwaia I (layer D)		Túnel I (Fourth Component)		Túnel I (Fifth Component)		Imiwaiwa I (layer B)		Shamakush I		Túnel I (Sixth Component)		Shamakush X	
	NISP	MNI	NISP	MNI	NISP	MNI	NISP	MNI	NISP	MNI	NISP	MNI	NISP	MNI	NISP	MNI	NISP	MNI	NISP	MNI	NISP	MNI
A.nigricans															1	1						
A. depressiceps	23	3	20	2	3	1	1	1			1	1	7	1	8	1	19	2	3	1	29	4
S. fueguensis	8516	190	1418	41					3	1					65	2	54	2	1	1		
C.gobio	9	4	182	6	5	1											3	1			11	2
E. maclovinus	15	3	364	9					1	1					1	1	1	1				
M. magellanicus	1600	61	1891	82	418	19	85	4	4	2	22	2	4	1	27	1	381	17	1	1	87	8
Merlucciidae	1432	40	3432	81	219	6	12	1	242	6	5	1	1	1	17	1	272	7	1	1	16	1
Merluccius sp.	11	4	439	16	2	1	2	1	1350	39	4	1	1	1	48	4						
Nototheniidae	236	11	413	14			2	1					1	1	6	1	38	13	34	3	7	1
P. magellanica	1196	37	1701	47					5	5	9	1	2	1	149	4	1022	30	91	4	191	6
Patagonotothen	349	11	67	3			3	1	13	1	3	1	3	2	25	1	603	21	102	2	11	1
S. australis	236	10	39	1			2	1	8	1												
Thyrsites atun	1	1	221	7											2019	28	212	3				
Totals	13624	375	10187	309	647	28	107	10	1626	56	44	7	19	8	2366	45	2605	97	233	13	352	23

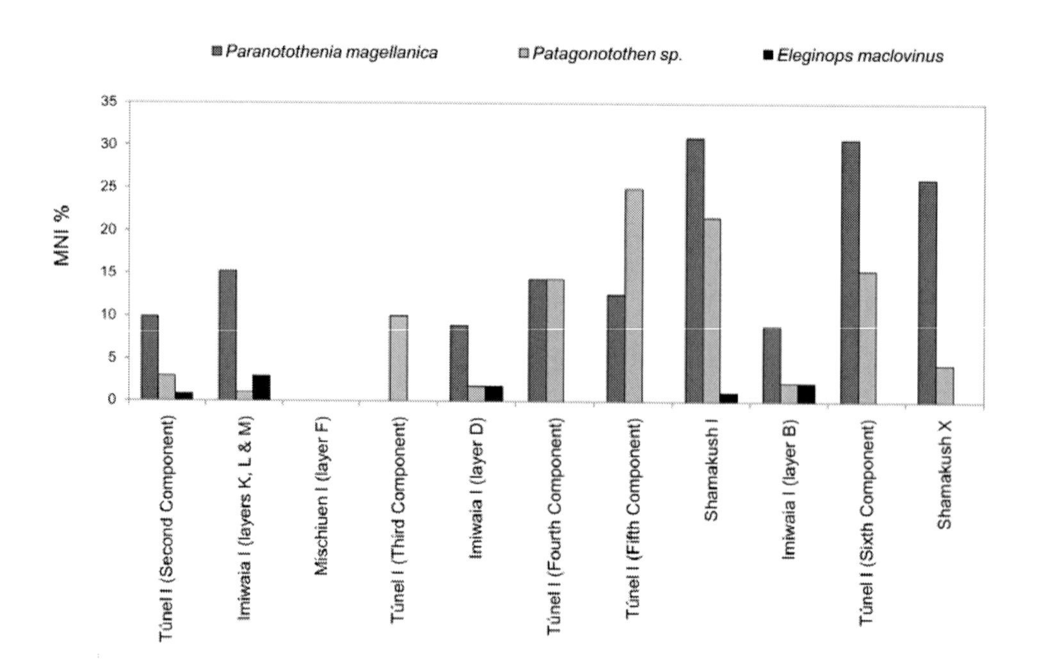

Figure 5. Relative abundances of *Paranotothenia magellanica, Patagonotothen* and *Eleginops maclovinus* throughout the archaeological sequence.

Merluciidae and Nototheniidae families are the mostly represented by their relative frequencies throughout the archaeological sequence of the Beagle Channel. In Merluciidae family, specimens attributable to *Macruronus magellanicus* appear in varying proportions (between 3% and 68% of MNI) in all assemblages analyzed (Figure 4). The *Merluccius* genus was recorded in eight assemblages: excepting Imiwaia I layer D, where the representation is almost 70% of the MNI, in other cases the frequencies are below to 15%. We also identified bone remains attributable to the Merluccidae family which could not be attributed to a more specific taxonomically level; these specimens are representations of MNI between 26% and 2%.

Among the nototheniids (Figure 5), *Paranotothenia magellanica* is the one with greater continuity in time sequence, with varying proportions of NMI between 31% and 9%. Specimens identified as *Patagonotothen* genus represent from 25% to 1% of MNI values. Bone remains corresponding to *Eleginops maclovinus* are present in 5 assemblages with representations of the MNI that do not exceed 3%. In several assemblages were also recovered remains attributable to Nototheniidae family which could not be determined at the genus or species level.

Some species identified in the archaeological record have proportionately high representations, but restricted to certain segments of the archaeological sequence (Figure 6). Representations of *Sprattus fueguensis* reach to 51% and 13% of MNI in the two earliest assemblages (Second Component of Túnel I and Layers K, L and M of Imiwaia I), and then this species appears at rates significantly lower in the rest of the sequence. Remains identified as *Salilota autralis* are presented in low proportions only in assemblages included in the period 6400 - 3000 radiocarbon years BP. By contrast, specimens of *Thyrsites atun* are represented in high proportions only in a late assemblage (layer B of Imiwaia I).

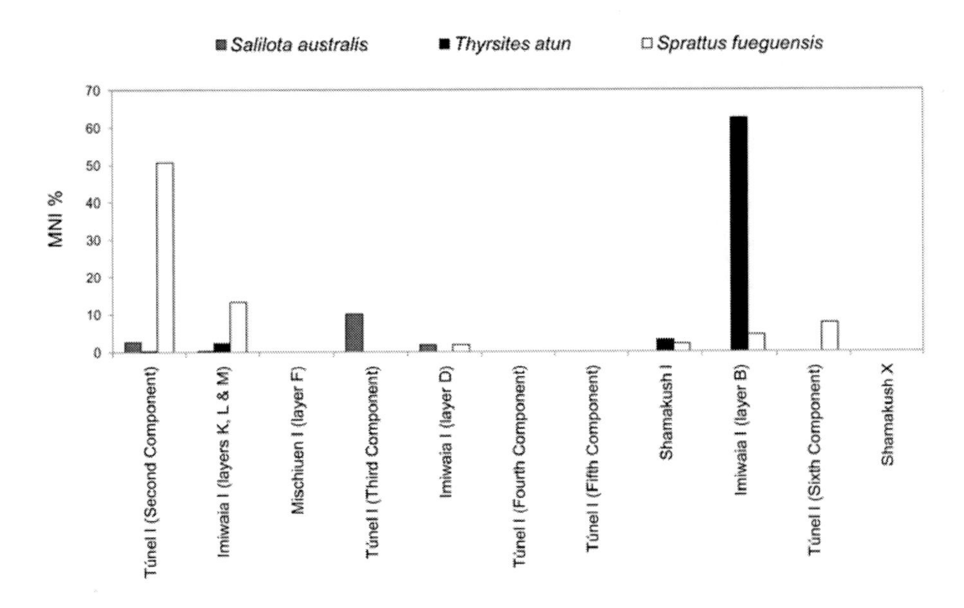

Figure 6. Relative abundances of *Salilota australis, Thyrsites atun* and *Sprattus fueguensis* throughout the archaeological sequence.

Other species are represented in lower proportions throughout the entire archaeological sequence (Figure 7). This is the case Zoarces as *Austrolycus depressiceps*, which is presented in 10 assemblages with proportions of MNI varying between 17% and 0.7%. *Cottoperca gobio* specimens are present at different moments of the sequence with frequencies lower than 9%. Finally, *Austroatherina nigricans* is represented by only one specimen in layer B of Imiwaia I.

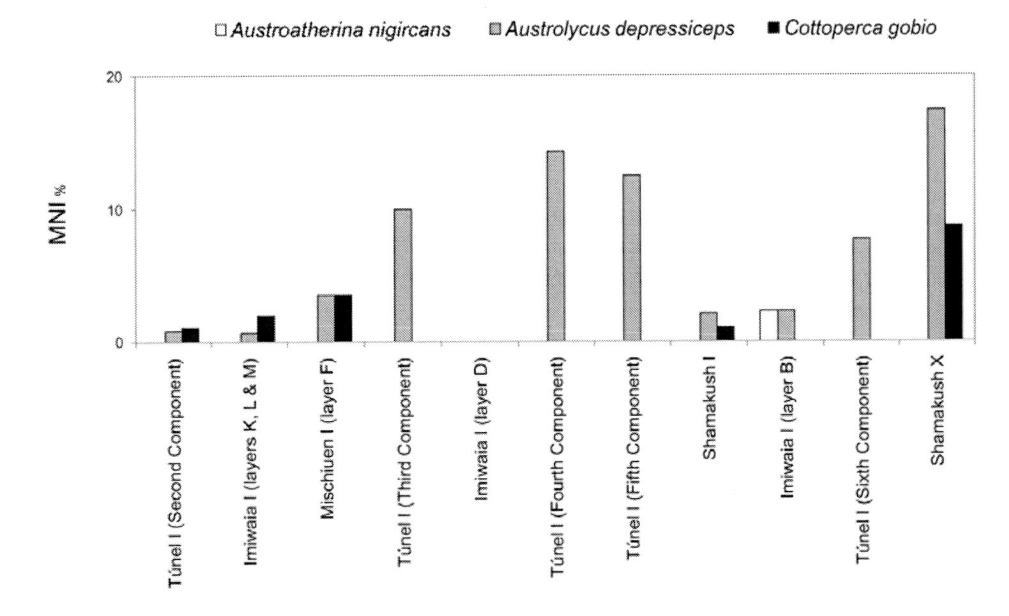

Figure 7. Relative abundances of *Austroatherina nigricans, Austrolycus depressiceps* and *Cottoperca gobio* throughout the archaeological sequence.

We assessed to what extent the composition of the ichtyoarchaeological records dependent of the degree of fragmentation and bone preservation. This concern arises mainly from the fact that many agents that cause bone loss or fragmentation (*v.g.*, overweight and compaction of sediments, diagenesis, etc.) may act differentially between the assemblages, affecting taxonomic representations mentioned above.

Table 4. Descriptive statistic for the overall averages of WMI% indexes for each of the analyzed assemblages

Archaeological assemblages	Ntaxa	Mean	Standard error	Lower limit	Upper limit
Túnel I (Second Component)	12	73,29	3,11	70,18	76,40
Imiwaia I (Layers K, L and M)	12	70,41	1,76	68,64	72,17
Mischiuen I (Layer F)	5	72,67	8,13	64,54	80,80
Túnel I (Third Component)	7	68,27	7,36	60,90	75,64
Imiwaia I (Layer D)	8	75,67	1,58	74,09	77,25
Túnel I (Fourth Component)	6	61,22	6,24	54,97	67,47
Shamakush I	10	77,46	2,08	75,37	79,55
Túnel I (Fifth Component)	7	80,18	2,65	77,52	82,84
Shamakush X	7	82,11	1,56	80,54	83,68
Túnel I (Sixth Component)	7	73,41	1,46	71,94	74,88
Imiwaia I (Layer B)	11	74,83	3,24	71,58	78,07

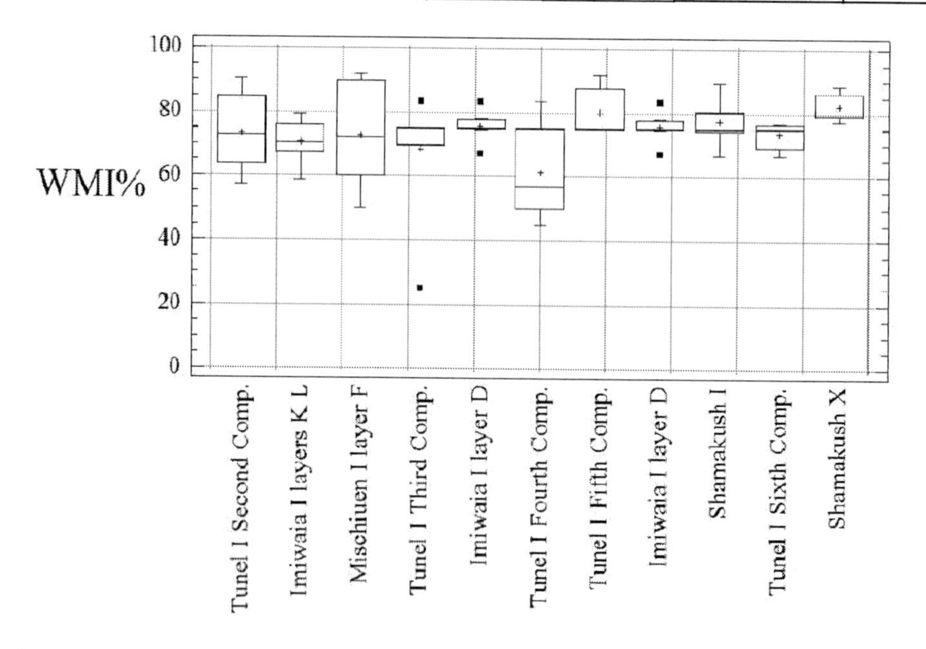

Figure 8. Box-plot of WMI% values throughout the archaeological sequence.

To address this issue, first we compare the overall average rates WMI% estimated for the different zooarchaological records. Table 4 shows means and standard errors for each assemblage, as well as the range of values using the minimum and maximum limits. We can observe that both early and late records maintained similar levels of bone integrity. The middle section of the sequence has greater variability, recording cases of high integrity as the Fifth Component of Túnel I, but also cases with high bone fragmentation, as the Fourth Component of the same site. Figure 8 shows the statistical information expressed through a box-plot diagram, which provides a more accurate picture of the spread of values in each assemblage. WMI% values are agglutinated in the range between 65% and 85%, showing little variability throughout the archaeological sequence. Therefore, the taxonomic variability of the ichtyoarchaeological assemblages does not seem to respond to differential preservation factors, allowing the discussion of other factors.

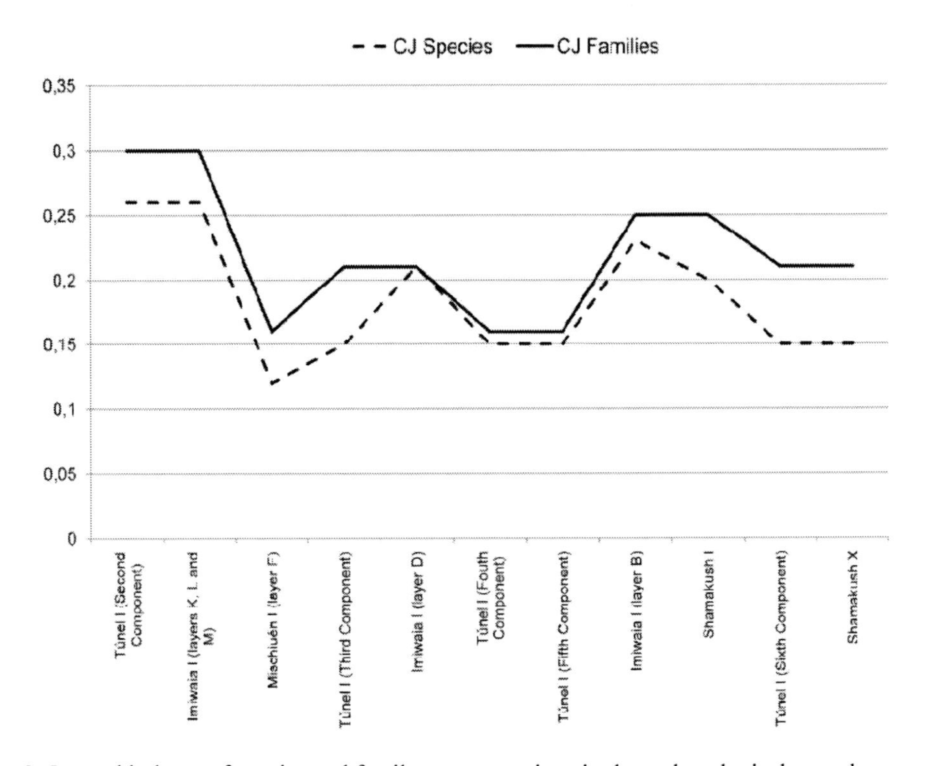

Figure 9. Jaccard indexes of species and family representations in the archaeological record.

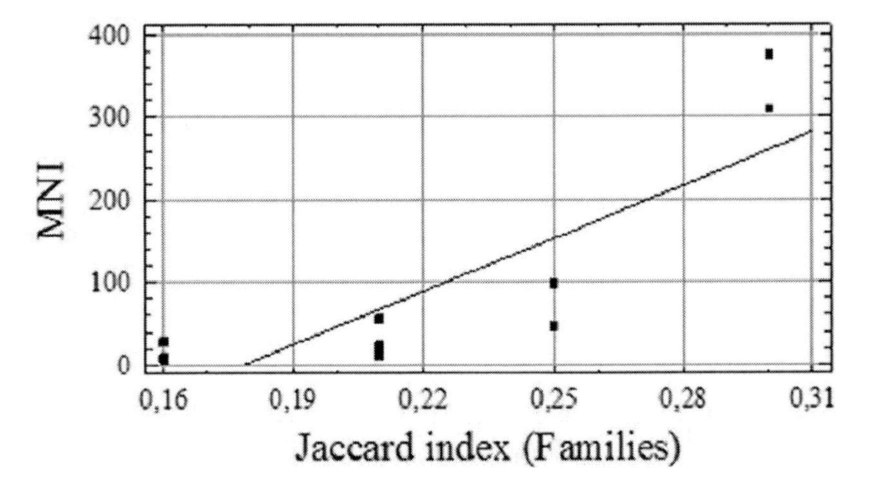

Figure 10. Correlation between the sample sizes and Jaccard indexes.

Discussion

Over a total of 20 families of bony fish currently recognized in the environment of the Beagle Channel, only 8 are represented along the archaeological sequence of that region. The difference at species level is even greater: over 45 fish bony species, a minimal of 11 species is represented in archaeological sites. This number of species could be greater as there are two genera identified (*Merluciius* and *Patagonotothen*) whose skeletal remains do not allow an osteological identification at more specific level.

The highest values of Jaccard indexes are recorded in the two earliest assemblages of the sequence, both at the family level as in the species level; this indicates a greater degree of similarity of these assemblages in relation to the current taxonomic composition of the Beagle Channel (Figure 9). Then there was a decrease in Jaccard indexes as we move through time to more recent occupations, which means that there is a greater distance between taxonomic representations of archaeological sites with respect to the current composition of fish fauna. However, if we evaluate the degree of correlation between these indices with the corresponding sample sizes, we can see that the Jaccard indexes in both family ($r = 0.84$ $p < 0.01$; Figure 10) and species ($r = 0.82$ $p < 0.01$) representations are affected significantly by the differences of sample sizes. Therefore, the trend observed may respond to a sampling problem and must be reevaluated in the future from the incorporation of new analysis.

While the faunal diversity recorded in zooarchaeological assemblages is low compared with the current composition of the ichthyofauna of the Beagle Channel, there are similarities and differences between the two sources of information. These variations stimulate the discussion of issues related to the distribution of fish fauna from the past to present condition.

Respect to those taxa with intertidal and coastal habits, along the entire occupational sequence there is a predominance of specimens attributable to *Paranotothenia magellanica*, *Patagonotothen* sp. and generally Nototheniidae family. With the exceptions of layer F of Mischiuen I and the Fifth Component Túnel I, these species constitute the most represented coastal resources in the zooarchaeological assemblages. Other taxa that live in intertidal and shallow waters as *Austrolycus depressiceps*, *Cottoperca gobio* and *Austroatherina nigricans* are represented in smaller proportions. These representations in the archaeological record do not contrast with modern records regarding the availability and abundance of fishes in coastal waters of the Beagle Channel. However, an unexpected factor is the low representation of *Eleginopsmaclovinus*. This species is presented by rare specimens throughout the archaeological sequence of the region, although in earlier times it recorded a relatively greater importance. This is important to point out because it is the species with higher biomass of the Beagle Channel (Lloris and Rucabado 1991, López *et al.* 1996). Moreover, this species is available throughout the year, making it difficult to explain its low representation in the archaeological record because of seasonal variations. Nor is there evidence to indicate that the abundance of *Eleginops maclovinus* was lower in the past than observed in modern records. Written sources by travelers, religious missionaries and ethnographers in the late 19 and early 20 centuries indicate that this fish resource could be captured in very abundant way (L. Bridges 1947: 95; Gusinde 1937: 534-535). The diversity of coastal species represented in the ichtyoarchaeological record of the Beagle Channel does not permit to assume that there were technological limitations of fishing*Eleginops maclovinus* (Estévez *et al.* 2001; Fiore and Zangrando 2006).In a previous study we interpret the low representation of *Eleginops* *maclovinus* as a possible consequence of social patterns of restriction, prohibition or avoidance (Fiore and Zangrando 2006) in the consumption of this resource. The idea that this could be a "taboo" seems likely to be propped by requirements implicit in the ideological sphere of hunter-gatherer societies of the region in ethnographic times (Fiore and Zangrando 2006).

While the absence or low representation of some coastal species in the archaeological record can be explained by socio-cultural factors (*v.g.Eleginops maclovinus*), the recurrence over the past 6400 years of species which inhabit in kelp forest of *Macrocystis pyrifera* confirms the importance of this ecosystem for the Beagle Channel area. These kelps provide shelter and food to many species of invertebrates and fish (Moreno and Jara 1984; Vanella *et al.* 2007), creating a space of trophic interactions between these animals, as well as with seabirds and marine mammals (Schiavini *et al.* 1997; Raya Rey and Schiavini 2000). Commercial use of *M. pyrifera* as a source of alginic acid has been considered an important potential resource for exploitation in Tierra del Fuego (Mendoza and Nizovoy 2000). The results presented in this chapter reinforces the warnings of biological studies about the indiscriminate exploitation of these kelps can significantly alter an important link in the ecology of the Beagle Channel (Mendoza and Nizovoy 2003; Vanella *et al.* 2007).

In the Second Component of Túnel I, the MNI attributable to *Sprattus fueguensis* is more than 50% over the total of individuals calculated, but this predominance is not repeated in other assemblages of the regional sequence. It is noteworthy that the presence of this resource in the stratigraphy of this component is bounded to the phases III, IV and V of layer D of this site, the contact between the latter two phases presents the highest density of remains, consisting of 7138 specimens. This grouping of bones has an area of about 1 m^2, where vertebrae and neurocranium fragments of a minimum of 143 individuals are represented. Therefore, the increased representation of *Sprattus fueguensis* at the beginning of the sequence would respond to a stochastic factor in which the product of a capture event was abandoned for some unknown reason.

Among the typical species of deep waters, *Macruronus magellanicus* is represented in the Beagle Channel archaeological sites at high proportions throughout the entire sequence. As we noted this species is typical of depths between 30 and 500 meters. However, in the Beagle Channel this species can be captured abundantly from shallow waters to 110 meters of depth, regardless of their size and maturity of the individuals (Lloris and Rucabado 1991). The frequency and distribution of this species appear to have been continuous in the marine environment for the past 6400 years; otherwise its conspicuous condition in the archaeological record would be difficult to explain.

Salilota australis is present only in early assemblages of the regional sequence, although not numerically significant in terms of subsistence.

Situations more difficult to explain from a hunter-gatherer technology and considering the current distribution of fish are the presence of certain species of deep waters in the archaeological record. Particularly two taxa -*Thyrsites atun* and *Merluccius*- have substantial representations in layers B and D of Imiwaia I site respectively. The record of *T. atun* in other sites of the Beagle Channel -that are not here analyzed- indicates that the layer B of Imiwaia I is not an isolated case, and the representation of this species clearly increased towards late moments of the regional archaeological sequence. On the one hand, in Lancha Packewaia site the presence of this species was identified in relatively high proportions only in layers with chronologies between 1500 and 300 radiocarbon years BP (Saxon 1979; Orquera and Piana

1993-94). Also in Lanashuaia I site, a hunter-gatherer occupation of the 19th century, the representation of *Thysites atun* exceeds 75% of the total of fish remains (Piana *et al.* 2000). Estévez and coauthors (2001) argued that the only possible explanation for this pattern has to be in an environmental factor.However, it is not easy to attribute the pattern of this resource to paleoclimate variations, since the presence of *Thysites atun* covers the entire period of the oscillations known as Medieval Climatic Anomaly and Little Ice Age, which are the records of the most extreme variation of the last 6400 years in the Beagle Channel (Obelic *et al.* 1998). Moreover, although in low proportions, this resource is also represented in early assemblages (Second Component of Túnel I and Layers K, L and M of Imiwaia I), so it was available for other times of the sequence. In previous analyses, one of us pointed out that the increase in archaeological representation of this species may have been related to changes in fishing strategies developed by groups of hunter-gatherers which inhabited the region (Zangrando 2009a, 2009b). This could have been linked to a widening of spatial ranges of fishing, which concerns to an increased in labor investment of this activity through more frequent incursions in offshore waters.But regardless of the causes that led to an increase in *Thysites atun* representations, the presence of this species in the archaeological record implies that it was abundantly available in the area of the Beagle Channel and that it was accessible through a hunter-gatherer technology, this implication contrastswith the modern records of the fishes.

It is noteworthy that in the total of sampling reported in Fenucci *et al.* (1974), Lloris and Rucabado (1991) and Lopez *et al.* (1996) *Thysites atun* was only recorded in one case on the eastern mouth of the Beagle Channel at 30 meters of deep. A similar situation can be indicated from the evidence from layer D of Imiwaia I, where *Merluccius* is represented by 70% of the total MNI.

Although with significantly lower proportions, it was also recorded a relative growth in the representation of this taxon in Fourth and Fifth Components of Túnel I. At the moment, it is difficult to define how we can project these trends over time, but again this species is not currently abundant in the Beagle Channel and its presence has only been documented at depths greater than 30 meters.

In sum, the fact that in the past both *Merluccius* and *Thysites atun* were exploited abundantly through hunter-gatherer technologies means that their distributions must have been different in the past. This does not necessarily imply that these resources were approaching the coasts at the level currently observed for *Macruronus magellanicus*, but they must have lived in pelagic waters at depths considerably lower. With respect particularly to the genus*Merluccius*, a factor that undoubtedly affected the distribution and abundance of this resource was the over-exploitation of fisheries in the Argentine Sea during the last decades. The overexploitation of fisheries resources is one of the most dramatic problems in the SW Atlantic. In mid-1970 the total catch of fish and invertebrates had reached 1 million metric tons, and the catch has been doubled since 1987. Since then, most important commercially species were *Merlucciushubbsi* and *Merlucciusaustralis*. These resources were overexploited regionally by mid-1990 (Esteves *et al.* 2000). The results obtained in this study allow to visualize from a broader temporal perspective the enormous impact generated by this process.

CONCLUSION

This chapter presented along-term analysis on the taxonomic diversity and distribution of fish faunaof the Beagle Channel. For this, taxonomic representations of zoo archaeological assemblages were taken into account, which were discussed in comparison with the current composition of the fish fauna. With regard to the fish ecology of the Beagle Channel we have arrived at two main conclusions:

a) the association of shallow-water taxa and closely linked to the ecosystem created by the forests of *Macrocystis pyrifera* was present for at least the last 6400 radiocarbon years, highlighting its importance in the trophic interactions along the evolutionary history of the region;
b) and distribution and abundance of deep waters species would have been different in the past, whichcouldbe particularly linked with the overexploitation of *Merluccius* developed in recent decades in the SW Atlantic.

ACKNOWLEDGMENT

We are very grateful to Angélica Tivoli, Ernesto Piana and Luis Orquera for their critical comments and suggestions. We thank Lucía Zangrando for their assistance in improving our use of English in the paper. This research was supported by CONICET (PIP 0395/10). Special thanks to the editor of this book (Sean P. Dempsey), who invited us to participate.

REFERENCES

Bridges, L. (1947 [1975]). *El último confín de la Tierra*. Marymar: Buenos Aires.

Esteves J.L., Ciocco N.F., Colombo J.C., Freije H., Harris G., Iribarne O., Isla I., Nabel P., Pascual M.S., Penchaszadeh P.E., Rivas A.L., and Santinelli N. (2000). The Argentine Sea: the southeast South American shelf marine ecosystem. In Sheppard C.R.C. (ed.), *Seas at the Millenium: An Environmental Evaluation* (749–771). New York: Pergamon.

Estévez, J., Piana, E.L., Schiavini, A., and Juan-Muns, N. (2001). Archaeological Analysis of the Shell Middens in the Beagle Cannel, Tierra del Fuego Island. *International Journal of Osteoarchaeology* 11:24-33.

Fenucci, J., Virasoro, C., Cousseau, M. B., and Boschi, E. (1974). Campaña Tierra del Fuego 74. Informe preliminar (Investigación pesquera). *Contribuciones del Instituto de Biología Marina* 261: 1-37.

Fiore, D, and Zangrando, A. F. (2006). Painted fish, Eaten fish: Artistic and archaeofaunal representations in Tierra del Fuego, Southern South America. *Journal of Anthropological Archaeology* 25: 371-389.

Grayson, D. K. (1984). *Quantitative zooarchaeology: topics in the analysis of archaeological faunas*. Orlando: Academic Press.

Gusinde, M. (1937[1986]). *Los Indios de Tierra del Fuego*. (Second Volume: Los Yámana). CAEA: Buenos Aires.

Iturraspe, R. J., and Schroeder, C. (1999). El clima en el canal Beagle. In *La vida material y social de los Yámana*: 36-45. Buenos Aires: Editorial Universitaria de Buenos Aires.

Jackson, J. B. C., Kirby, M. X., Berger, W. H., Bjorndal, K. A., Botsford, L. W., Bourque, B. J., Bradbury, R. H., Cooke, R., Erlandson, J., Estes, J. A., Hughes, T. P., Kidwell, S., Lange, C. B., Lenihan, H. S., Pandolfi, J. M., Peterson, Ch. H., Steneck, R. S., Tegner, M. J., and Warner, R. R.. (2001). Historical Overfishing and the Recent Collapse of Coastal Ecosystems. *Science* 27 (July): 629-637.

Jones, S. (2007). Human Impacts on Ancient Marine Environments of Fiji's Lau Group: Current Ethnoarchaeological and Archaeological Research. *The Journal of Island and Coastal Archaeology* 2(2): 239-244.

Leible, M., Alveal, E., and Maldonado, J. (1981). *Catálogo de peces que habitan las aguas costeras de la bahía Concepción y bahía de San Vicente*. Talcahuano: Biotecmar, Universidad Católica de Chile.

López, H. L., García, M., and San Román, N. (1996). Lista comentada de la ictiofauna del Canal Beagle, Tierra del Fuego, Argentina. Ushuaia: CADIC, Contribución Científica (Special Issue).

Lyman, R. L. (1996). Applied zooarchaeology: The relevance of faunal analysis to wildlife management. *World Archaeology* 28: 110-125.

Lyman, R. L. (2006). Paleozoology in the Service of Conservation Biology. *Evolutionary Anthropology* 15: 11–19.

Lyman, R. L., and Cannon, K. P. (Editors). (2004). *Zooarchaeology and Conservation Biology*. Utah: University of Utah Press.

Lloris, D., and Rucabado, J. (1991). Ictiofauna del canal Beagle (Tierra del Fuego): aspectos ecológicos y análisis biogeográfico. Madrid: Instituto Español de Oceanografía (Special Issue 8).

Magurran, A. E. (1989). *Diversidad ecológica y su medición*. Barcelona: Vedrá.

Mendoza, M.L., and Nizovoy, A. (2000). Géneros de macroalgas marinas de la Argentina, fundamentalmente de Tierra del Fuego. Poder Legislativo de la Provincia de Tierra del Fuego, Antártida e Islas del Atlántico Sur, Ushuaia.

Mendoza, M.L., and Nizovoy, A. (2003). Desarrollo sustentable de los recursos acuáticos vivos del Canal Beagle. Respuesta a cortes experimentales de un bosque de *Macrocystis pyrifera* (Cachiyuyo) del Canal Beagle. Informe final. Gobierno de Tierra del Fuego, Antártida e Islas del Atlántico Sur Ushuaia.

Moreno, C. A., and Jara, H. F. (1984). Ecological studies on fish fauna associated with *Macrocystis pyrifera* belts in the south of Fueguia Islands, Chile. *Marine Ecology-Progress Series* 15: 99-107.

Obelic, B., Álvarez, A., Argullós, J., and Piana, E. L. 1998. Determination of water palaeotemperature in the Beagle Channel (Argentina) during the last 6000 years through stable isotopic composition of *Mytilus edulis* shells. *Quaternary of South America and Antarctic Peninsula* 11: 47-71.

Orquera, L. A., and Piana, E. L. (1993-94). Lancha Packewaia: actualización y rectificaciones. *Relaciones de la Sociedad Argentina de Antropología* XIX: 325-362.

Orquera, L. A., and Piana, E. L. (1999a). *La vida material y social de los Yámana*. Buenos Aires: Editorial Universitaria de Buenos Aires.

Orquera, L. A., and Piana, E. L. (1999b). *Arqueología de la región del canal Beagle (Tierra del Fuego, República Argentina)*. Buenos Aires: Sociedad Argentina de Antropología.

Piana, E. L., Estévez Escalera, J. and Vila Mitjá, A. (2000). Lanashuaia: un sitio de canoeros del siglo pasado en la costa norte del canal Beagle. In *Desde el País de los Gigantes. Perspectivas arqueológicas en Patagonia,* Volume II: 455-469. Río Gallegos: Universidad Nacional de la Patagonia Austral.

Rabassa, J., Heusser, C. and Stuckenrath, R. (1986). New data on Holocene sea transgressión in the Beagle Channel (Tierra del Fuego). *Quaternary of South America and Antarctic Peninsula* 4: 291-309.

Rae, G. (1991). Biología reproductiva comparada de dos especies de nototénidos del Canal Beagle, Argentina. PhD dissertation, Universidad Nacional de La Plata, La Plata.

Raya Rey, A., and Schiavini, A. C. M.. (2000). Distribution, abundance and associations of seabirds in the Beagle Channel, Tierra del Fuego, Argentina. *Polar Biology* 23:338–345.

Reitz, E. J. (2004). The Use of Archaeofaunal Data in Fish Management. In R. C.G.M. Lauwerier and I. PLUG (eds.),*The future from the past: archaeozoology in wildlife conservation and heritage management* (19-33). Oxford: Oxbow.

Reitz, E.J., Quitmyer, I. R., and Marrian, R.A. (2009). What Are We Measuring in the Zooarchaeological Record of Prehispanic Fishing Strategies in the Georgia Bight, USA? *Journal of Island and Coastal Archaeology* 4: 2-36.

Reitz, E., and Wing, E. (1999). *Zooarchaeology*. Cambridge: Cambridge University Press.

Saxon, E. C. (1979). Natural Prehistory: the archaeology of Fuego-Patagonian ecology. *Quaternaria* XXI: 329-356.

Schiavini, A.C.M., Goodall, R.N.P., Lescrauwaet, A.K., and Alonso, M.K. (1997). Food habits of the Peales Dolphin, *Logenorhynchus australis*; review and new information. *Rep. Int. Whale Commun.* 47:827–834.

Vanella, F., Fernández, D. A., Romero, M. C., and Calvo, J. (2007). Changes in the fish fauna associated with a sub-Antarctic *Macrocystis pyrifera* kelp forest in response to canopy removal. *Polar Biology* 30:449–457.

Zangrando, A. F. (2009a). *Historia evolutiva y subsistencia de cazadores-recolectores marítimos de Tierra del Fuego*. Buenos Aires: Sociedad Argentina de Antropología.

Zangrando, A. F. (2009b). Is fishing intensification a direct route to hunter-gatherer complexity? A case study from the Beagle Channel region (Tierra del Fuego, southern South America). *World Archaeology* 41 (4): 589 – 608.

Zohar, I., Dayan, T., Galili, E., and Spanier, E., (2001). Fish processing during the early Holocene: A taphonomic study. *Journal of Archaeological Science*: 28: 1041-1053.

In: Fish Ecology
Editor: Sean P. Dempsey

ISBN 978-1-61324-282-7

Chapter 8

ECOLOGY OF EARLY LIFE-HISTORY STAGES OF ANADROMOUS SHADS

Eduardo Esteves*

Centro de Ciências do Mar CCMar – CIMAR Laboratório Associado,
Campus de Gambelas, Faro and Instituto Superior de Engenharia da Universidade do Algarve, Campus da Penha, Faro

ABSTRACT

Shads (Clupeidae: subfamily Alosinae) are a cosmopolitan group of fishes that exploit a wide range of habitats throughout the world, occurring in lakes, rivers, and seas. They are valuable ecological and economical resources. Species are able to adapt to estuarine, lentic and/or lotic habitats and present large plasticity in reproductive features, *e.g.* there is evidence of hybridization between allis shad *Alosa alosa* and twaite shad *A. fallax*. On the other hand, shads constitute important recreational and commercial fishes, *e.g.* American shad *A. sapidissima* in North America, allis shad in Europe or Indian shad *Tenualosa ilisha* in the Indo-Pak subcontinent and the Persian Gulf region. Most of these alosines are anadromous (a few landlocked populations have been found in Portugal, Ireland, Eastern Europe or Northern India), some species migrating several hundred kilometers upriver to spawn (*e.g.* 800 km in the case of allis shad or as much as 1200 km in the case of Indian shad), and others exhibiting a pronounced homing behavior similar to that of migratory salmonids (*e.g.* allis shad). Adults are usually fished during the upriver movements towards spawning grounds but catches have been declining worldwide during the last 20 to 30 years, mostly due to anthropogenic activity, *e.g.* damming of rivers, overfishing (the predictability of migrations has rendered them vulnerable to overharvest) and deterioration of habitats by industrial and agricultural pollution.

Environmental conditions in the freshwater/brackish water reaches of rivers during the embryo-larval period are thought to play an important role in the future of populations, purposely determining the recruitment variability that is characteristic of the

* Centro de Ciências do Mar CCMar – CIMAR Laboratório Associado, Edifício 7, Campus de Gambelas, Faro & Instituto Superior de Engenharia da Universidade do Algarve, Campus da Penha, 8005-139 Faro. E-mail: eesteves@ualg.pt

alosines. Moreover, the successful domestication of shads, American shad in USA and Reeves shad *T. reevesii* in China, and use as broodstock for restoration programs largely depends on understanding the biology of eggs, larvae and juveniles. Notwithstanding, the early life-history stages have been relatively poorly studied in European, Middle Eastern or African shads as opposed to North American species. A decade ago, two international conferences held in Bordeaux (France), in 1999, and in Baltimore (USA), in 2001, assembled the knowledge on world shads, namely their biology. Since then, relevant work on early life-history stages of shads worldwide has been carried out. Herein, I compile, update and integrate these recent contributions on the biology and ecology of egg, larval and juvenile stages of shads, and prospect future work.

INTRODUCTION

Shads (Clupeidae: subfamily Alosinae) are a cosmopolitan group of diadromous fishes that exploit a wide range of habitats worldwide, occurring in lakes, rivers, and seas.

Alosines are valuable ecological resources; species are able to adapt to estuarine, lentic and/or lotic habitats and present large plasticity in reproductive features, *e.g.* there is evidence of hybridization between allis shad *Alosa alosa* and twaite shad *A. fallax fallax* [1, 2].

Of the 31 species in seven genera (*Alosa, Brevoortia, Ethmalosa, Ethmidium, Gudusia, Hilsa* and *Tenualosa*) included in the subfamily, half are anadromous (genus*Alosa, Hilsa* and *Tenualosa*) and the remaining are marine (menhadens, genus *Brevoortia* or *Ethmidiummaculatum*), amphidromous (e.g. *A. brashnikovi*, Bonga *Ethmalosa fimbriata*, or Longtail shad *Tenualosa macrura*) and freshwater riverine (Chapra and Burmese shad, genus *Gudusia*) [3, 4]. During the anadromous migration they can cover large distances from marine to freshwater habitats, e.g. 800 km in the case of allis shad [3, 5, 6] or about 1200 km in the case of Indian shad or hilsa *T. ilisha*[7]. Recently, Limburg [8] demonstrated that the migratory movements of diadromous fishes are far more complex than previously thought. Some anadromous alosines have become landlocked in Spain, Portugal and Morocco[9]; others are invasive species in lakes, e.g. alewife *A. pseudoharengus* in Lake Michigan, using various spawning and nursery habitats: near shore sites, drowned river-mouth lakes, bays, etc. [10]. Moreover, some lacustrine populations have been considered as species, e.g. *A. macedonica* (Greece), while others have been given subspecies status, e.g. *A. fallax lacustris* (Italy) and *A. f. killarnensis* (Ireland). Blaber et al.[11] questions the classification of Indian shad as truly anadromous since it is very salinity tolerant and inhabits freshwater, estuarine and coastal waters in the Bay of Bengal.

Evidence to date indicates that anadromous alosines home well to their natal rivers [4, 12]. Using geochemical signatures in otoliths and water collected from 20 major spawning rivers along the Atlantic coast of USA and Canada, Walthers and Thorrolds[13] were able to discriminate the natal origin of American shad *A. sapidissima* with classification accuracies of 93%. Similar levels of accuracy (above 85%)were attained by Tomás et al [6] when using otoliths of allis shad. Seemingly, homing is a mechanism that promotes the development and persistence of stocks [14] and allows genetic differences to accumulate [15]. In contrast, most hilsa in Bangladesh returned to freshwater after reaching sexual maturity, but not necessarily to their natal river [16]. Olfaction is thought to play a major role in the migratory movements towards natal rivers, namely for American shad [17].

This stock-river relationship is important considering that environmental conditions during the embryo-larval period in the freshwater reaches of rivers play an important role in the future of populations [18]. Crecco and Savoy[19] showed that American shad year-class strength depends upon parent stock size and environmental conditions during the larval stages. Habitat loss (especially damming), overfishing, pollution, and, increasingly, climate change, nonnative species, and aquaculture contributed to the dramatic declines in diadromous fishes in North Atlantic [20, 21]. Gradually, the declines may also lead to the loss of institutional and societal memory about past abundance and importance, contributing to lack of motivation and funding for restoration programs[21]. Conversely, other populations have proven robust despite barrages, pollution, exploitation, and environmental changes, e.g. the critical DanubeRiver population of Pontic shad *A. pontica*[22].

In contrast to salmonids shads are more sensible to the presence of obstacles (dams, weirs, etc.) in their migratory pathways because shads are incapable of jumping and cannot swim for long periods, thence fish cannot access natural, historical spawning sites. These barriers pose a threat to the survival of shad populations, *e.g.* in <20 years allis shad and twaite shad populations disappeared from the River Douro (Portugal) as a consequence of the construction of several dams in the estuary [23]. In recent years, different fish passes have been developed and implemented for allis and twaite shad, notwithstanding there has been very limited success – 10 to 20 % efficiency are common figures [cf. 24]. In contrast, the successful restoration of depleted populations of the alosines in North Americahas been carried out for several decades through transplanting of sexually mature, prespawn adults or stocking of cultured larvae and juveniles[25].

Shads constitute important recreational and commercial fishes that are harvested throughout their range in estuaries, middle sections of rivers or in their resident form in lakes. Notwithstanding, fisheries have been focused mainly on shad when they migrate towards the spawning grounds, since during the highly predictable upstream runs specimens naturally aggregate in specific locations. Despite being an important resource, the economic value of the catches of allis shad has been seriously reduced as a result of the decline or collapse of the stocks [26] and many populations of American shad are under moratorium[27] and stock abundances throughout their range are at historic lows [28]. Once the major commercially fished species in USA, American shad catches declined rapidly from an historic high of ca. 23,000 metric tons (in 1896) to present-day levels in the hundreds of metric tons. Similarly, native alewife and blueback herring populations on the East coast of USA have suffered declines in abundance in the past 2-3 decades, from 25,000 metric tons during 1950-1969 to the current level of 1,000 metric tons [29]. The trends in the annual catches of allis and twaite shad in 1978-1999 for various river systems are shown in (Figure 1). In the Yangtze, Pearl and Qiantang rivers (China), the populations of Reeves shad *T. reevesii* supported a lucrative commercial fishery before de 1960s; after the crash in 1978, populations are in the verge of extinction[30]. Indian shad (or hilsa) is an important resource in the Indo-Pak subcontinent and the Persian Gulf region, wherein it is the basis of very large fisheries in Burma, Bangladesh and India, totaling well over 200,000 metric tons per year[11]. The fishing of juvenile hilsa, using bagnets, is commonly observed from November to May (occasionally up to July) in a freshwater tidal section of the Hoogly estuarine system (India) and its upper freshwater riverine zones [31]. According to CIFRI [31], a 50 % reduction in the fishing mortality of juvenile hilsa, with an estimated catch that fluctuated between 41 – 150 metric tons per year (1998-2003), would have the potential to augment 10% the adult production.

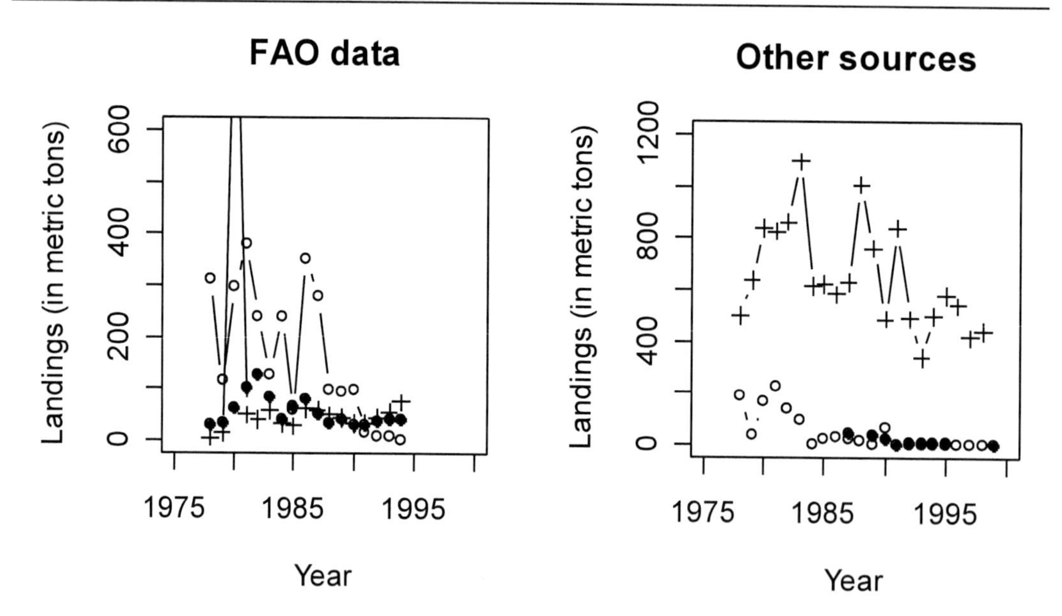

Figure1. Annual catches of Allis and Twaite shads reported to (left) the Fisheries Department of the FAO by national institutions of France (+), Portugal (•) and Morocco (○) and (right) compiled for key rivers from several other sources [3, 32, 33, 34]: Gironde-Garonne-Dordogne rivers system in France (+), River Lima in Portugal (•) and River Sebou in Morocco (○).

The amount of raw or lightly treated human sewage can act as a "chemical dam" due the induced low levels of oxygen[21]. Maes etal [35] found evidence of that effect on twaite shad just upstream of the freshwater-saltwater boundary in River Scheldt (Belgium). Those authors derived a statistically significant logistic regression model describing the effects of dissolved oxygen (DO), temperature (T) and flow (Q) upon the presence of twaite shad:

$$\text{logit}(p) = -10.85 + 0.77DO - 0.014Q + 0.67T - 0.014T^2 + \varepsilon .$$

Interestingly, the probability to capture adult shad increases with increasing dissolved oxygen (given temperature and flow), but decreases at higher river flow rates (given temperature and DO).

Climate change is altering species distribution. Warming appears to be shifting the phenologies of anadromous fishes towards earlier spawning runs [21]. Impacts of climate change in the Loire River (France) thermal and hydrological regimes in the 1976-1995 period influenced the departure of young-of-the-year (YOY)allis shad on their downstream migration(by 17 days) but not their annual abundance [36].

All these attributes make alosines a privileged indicator-species of the biological (and physical) quality of the lower and middle reaches of rivers [3, 37].

The impact of human activities has led to a drastic restriction and fragmentation of the distribution area of shads. Currently, the distribution of allis shad and twaite shad as regressed to limited areas, particularly in the case of allis shad. Nowadays, this species only occurs in rivers along the Atlantic coasts of France and Portugal[3, 9] when, historically, allis shad occurred along the Atlantic eastern coast, from Norway to Morocco, extending along the British Isles, the coasts of Germany, Belgium, France, and then southward to Spain and Portugal [see the references in 3]. The contraction of the geographical range of several species of *Tenualosa* spp. and the large decline in their fisheries mirror what has happened to other alosines worldwide. Declines in catches of hilsa in the Shatt al Arab estuary (Iraq-Iran border) have been attributed to diking, damming, and draining of wetland areas that had in impact on shad spawning and nursery grounds [38]. Moreover, Reeves shad and Laotian shad *T. thibaudeaui* are on the brink of extinction in South and Southeast Asia[11]. In contrast, despite the dramatic declines in fisheries catches, American shad have, however, become widespread in several river systems following introduction/transplanting efforts in the last decades [39].

In many countries, alosines are now either extinct or are regarded as rare or endangered according to IUCN criteria (Table I) [9, 26]. The few attempts of artificial propagation, either to restore or augment population's abundance and distribution, of *Alosa* spp. in Europehave been unsuccessful[3] whereas experimental results show that the artificial culture of Reeves shad in China is more than just feasible [30, 40] and in North America (especially in USA) culture, stocking and translocation of *Alosa* spp. is a common and successful practice [cf. 25]. Only recently, a four-year, multi-funded, large-scale project successfully cultured and stocked five million allis shad larvae in the River Rhine (Germany, Holand, France)[1]. The successful domestication of shads and use as broodstock for restoration programs largely depends on understanding the biology of eggs, larvae and juveniles. Notwithstanding, the early life-history stages have been relatively poorly studied in European, Middle Eastern or African shads as opposed to North American species [e.g. 19, 41-51]. Surely, one reason is that early life-history stages are difficult to collect.

Table I. Current level of species conservation status of anadromous shads worldwide

Common name	Scientific name	Conservation status	Source
Agone	*Alosa agone*	LC	[52]
Alabama shad	*Alosa alabamae*	DD, SC	[52], [53]
Alewife	*Alosa pseudoharengus*	SC	[53]
Allis shad	*Alosa alosa*	LC	[52]
Killarney Shad	*Alosa killarnensis*	CE	[52]
Macedonian Shad	*Alosa macedónica*	VU	[52]
	Alosa vistonica	CE	[52]
American shad	*Alosa sapidissima*	Lowest in history	[54]
Blueback herring	*Alosa aestivalis*	SC	[53]
Caspian shad	*Alosa caspia*	LC	[52]
Hickory shad	*Alosa mediocris*	Status unknown	[53]
	Tenualosa thibaudeaui	EN	[52]
Twaite shad	*Alosa fallax*	LC	[52]

CE – critically endangered; DD – data deficient; EN – endangered; LC – least concern; VU – vulnerable; SC – species of concern.

A decade ago, two international conferences held in Bordeaux (France) in 1999 [55], and in Baltimore (USA) in 2001 [56], assembled the knowledge on world shads, namely their biology and ecology. Since then, relevant work on early life-history stages of shads worldwide has been carried out. In this chapter, I compile, update and integrate these recent

[1]More information at http://www.lanuv.nrw.de/alosa-alosa/en/index.html, accessed on February 19, 2011.

contributions on the biology and ecology of egg, larval and juvenile stages of anadromous shads, and prospect future work.

Spawning Migration and Habitat Requirements for Eggs

In anadromous *Alosa* spp., mature individuals (3 to 6 years-old depending on species and habitat) migrate from growing marine areas to spawning areas upriver during a period of few months, between December and August, and along a latitudinal gradient from South to North in a temporally progressive manner [9, 26, 50, 56]. The relative fecundity of shads varies with species, habitat and estimation methodology but oocytes counts or ovules per kg weight usually reach hundreds of thousands. Olney and McBride [57] did not confirm the hypothesis of Legget and Carscadden [58] of reciprocal latitudinal trends in relative fecundity and degree of repeat spawning (iteroparity) in reproductively isolated populations of anadromous American shad across its native range.In Bangladesh, hilsa spawns throughout the year with peak periods that vary among areas. Nonetheless, there seems to exist two seasons of increased spawning activity, one during the monsoon (July-October) and a second in January-March [11]. Rahman [7] reduces the period of peak spawning for hilsa in Bangladesh to September-October. Hilsa spawns in rivers, in estuaries and, to lesser extent, on the coast [11].

The pattern of migratory dynamics of allis shad corresponds to a dense anadromous migration of fish with one or two peaks of migration [59] whereas twaite shad enters the river in a series of waves [60]. There is a higher proportion of males at the start of the migration [see 26 for a review] but this unbalanced sex ratio tends to level out by the end of the spawning season [59].American shad are serial, broadcast spawners [48], i.e. capable of spawning several times during a single season. The same behavior applies to 25-38% of the adult Alabama shad*A. alabamae* in the Apalachicola River (USA) [50]. Generally, anadromous allis shad are considered semelparous since no more than 5-6% of the Atlantic stock spawn more than once [2]. Conversely, nearly all populations of twaite shad have an iteparous life history[26].

Water temperature, which has a strong influence on hydrology and ecology both in estuaries and freshwater, is considered the most important environmental factor moderating migratory behavior. Spawning migrations of allis shad and twaite shad generally take place when river temperature ranges from 12 to 20 °C [9, 26], a shorterrange than that reported for American shad, 8 to 26°C [28]. Spawning of Alabama shad occurs at relatively higher temperatures, 19-23 °C [61]. Other environmental factors like tidal rhythm, water discharge turbidity and salinity are also deemed preponderant, namely for allis and twaite shads [59]. The temperature thresholds assure that ambient temperature is appropriate for egg development.

Allis shad spawns during the night at sites typically located in the middle or upstream reaches of the river. Spawning of twaite shad has been reported in tidal freshwater[62, 63] as well as in the nontidal freshwater areas of rivers [64] during the night. Some authors have narrowed down the maximum spawning activity of allis shad, twaite shad and Mediterranean shad *A. f. rhodanensis* to overnight between 22:00 and 03:00[65]. A commotion caused by tail splashing and rapid circular swimming near the surface (also known as "bull") is characteristic of spawning behavior of allis shad [9, 66, 67]. Courtship and spawning behavior of American shad consists of running together, circling, splashing, parallel swimming, and gamete release and a there exists correspondence between audible splashing and the appearance of fertilized eggs in the water [68 and references therein].

The typical spawning habitat of allis shad in France is characterized by an area of coarse substrate limited upstream by a pool and downstream by shallow water with fast-moving currents [65], whereas the spawning habitat of twaite shad in UK rivers comprises a fast-flowing shallow area of unconsolidated gravel-pebble or cobble substrate of depths <1.0 m[64]. Commonly, however, there is an area of deep water or a deep pool, where adults accumulate during the day, close to shallow, sandy-gravel rifle areas where they move onto to spawn during the night [66, 69]. In the River Elbe (Germany), twaite shad is reported as spawning in the upper reaches of the estuary in depths of up to 8.0-9.5 m[62]. Esteves and Andrade [70] only collected early life-history stages of twaite shad in upstream, riverine sites in the River Mira (Portugal) located close to the upper boundary of estuarine influence and in the vicinity of the suspected spawning grounds. At those sites, river width averaged 30 m, depth<3 m and substrate consisted of a mixture of plain mud/silt, near the margins, and coarse-gravel or even stones, in the main channel, together with submerged vegetation and debris, river flow <1.5 m s^{-1} and water salinity<5 ppt. American shad are known to broadcast their eggs at depths of 1 to 10 m, and prefer sand, gravel or a mixture of both substrates[28]. The closely-related Alabama shad prefers to spawn over sandy substrates or near sandbars in moderate current downstream of dams, gravel shoals or in confined sites [61] when water temperatures range from 10 to 22 °C [50].

After spawning, the eggs of most *Alosa* spp. (1-4 mm in diameter before hydration) drift in the current before sinking to the bottom where they become embedded in small crevices in the substrate. In contrast, the eggs of Pontic shad remain pelagic until hatching [71]. Environmental conditions, mainly temperature, regulate the rate of embryonic development. Although the precise (negative) relationship is species-specific, over a broad range of temperatures the simple power function

$$I = \alpha \cdot T^{\beta}$$

(where α and β are empirical constants) has been applied [72]. Eggs successfully develop at water temperatures above 17°C for allis shad, incubating for 4-8 days[9], and between 15 and 25°C for Twaite shad with incubation taking 72-120 hours [60, 73]. Limburg [44] found that egg development time (EDT, days) for American shad can be expressed as function of temperature (*T*, °C):

$$\ln(EDT) = 8.9 - 2.484 \cdot \ln(T)$$

This equation, however, overestimates the duration of egg development in other alosines particularly at lower temperatures, e.g. in twaite shad at 15 °C the incubation is referred to last 72 h (or 3 d) but the model predicts 6.2 d (148.6 h), whereas at 25 °Cthe incubation duration is 120 h, or 5 d, and the equation above predicts 136 h (5.6 d).

LARVAL SHAD

Hatching of American shad occurs over a wide range of environmental conditions during a protracted period. The limit of early hatch dates is determined by flow and temperature (environmental conditions) and the limit of late hatching is determined by size at and timming of emigration[48].

The size (length) of freshwater larvae at hatching is considerably greater than that of marine larvae [72]. Larvae of allis shad are 7-12 mm TL at hatch but twaite shad larvae are relatively smaller at hatch, 5-8 mm TL [74]. Similarly, the size of American shad at hatching ranges from ca. 6.5 to 10 mm TL. Indian shadand Reeves shad hatchlings are much smaller than the abovementioned alosines, 2.3 to 3.1 mm[7] and 2.69 mm [30] in length respectively.

After hatching, allis shad larvae move to open water, exhibiting positive photoresponse[9, 75, 76] and adopting a nektonic behavior that persists until they are about 30 days-old [77]. Likewise allis shad, positive response to lightshould also be expected in American shad since the two species are closely related even if they are geographically isolated [37]. From field sampling, Esteves and Andrade [70] found evidence of positive photoresponse in larvae of twaite shad but not for yolk-sac larvae (Figure 2). Yolk-sac larvae of American shad in the middle-to-upper Delaware River were found to be more numerous after 18:00h whereas larvae densities were similar over all hours sampled, from 09:00 to 01:00[68]. Conversely, Gadomski and Barfoot[78] collected significantly greater proportions of yolk-sac larvae and larvae of American shad at night in the Columbia River (USA). Seemingly, the diel distribution of shads can vary depending on ecosystems, species and possibly years.

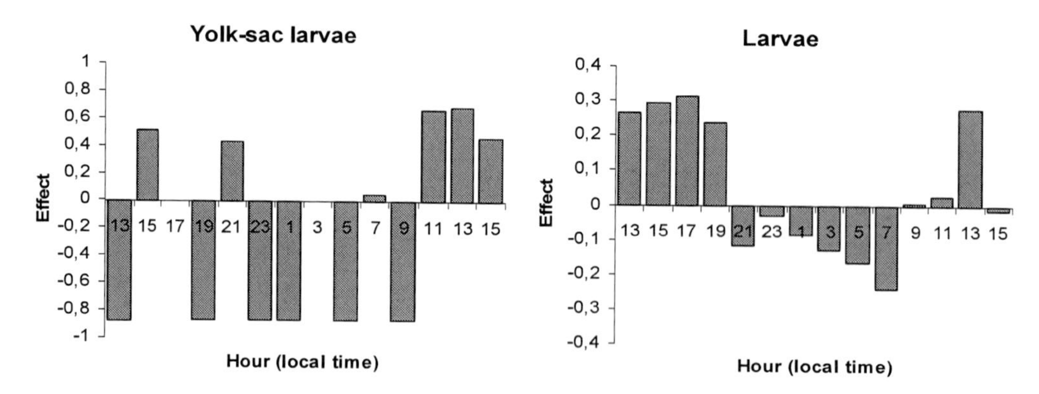

Figure 2. Effects of hour of sampling on yolk-sac larvae and larval density (note the different y-axis scales). The results of a Median Polish Analyses are scaled to a median of zero, thus positive values denote more specimens than expected from a random distribution; conversely for negative values. Approximate hours of sunrise and nightfall were 06:00 and 20:00 (redrawn from Esteves and Andrade [70]).

The nektonic behavior of shad larvae makes individuals more susceptible to undergo downstream drift. Larvae of both allis shad and twaite shad prefer low currents areas (*e.g.* side-channels or backwaters) which provide better nursery and feeding areas [73, 74]. After becoming nektonic, American shad larvae (8-25 mm TL) seemingly move to shorelines in the Columbia River (USA) [39]. Moreover, larvae of hilsa prefer marginal waters in the regions were the water depth does not exceed approximately 1.3 m[79]. In the highly turbid Danube River (Romania, Ukraine), Pontic shad larvae drift passively to the Black Sea in the middle of the strong water current and in the upper 0.5 mof water [71].

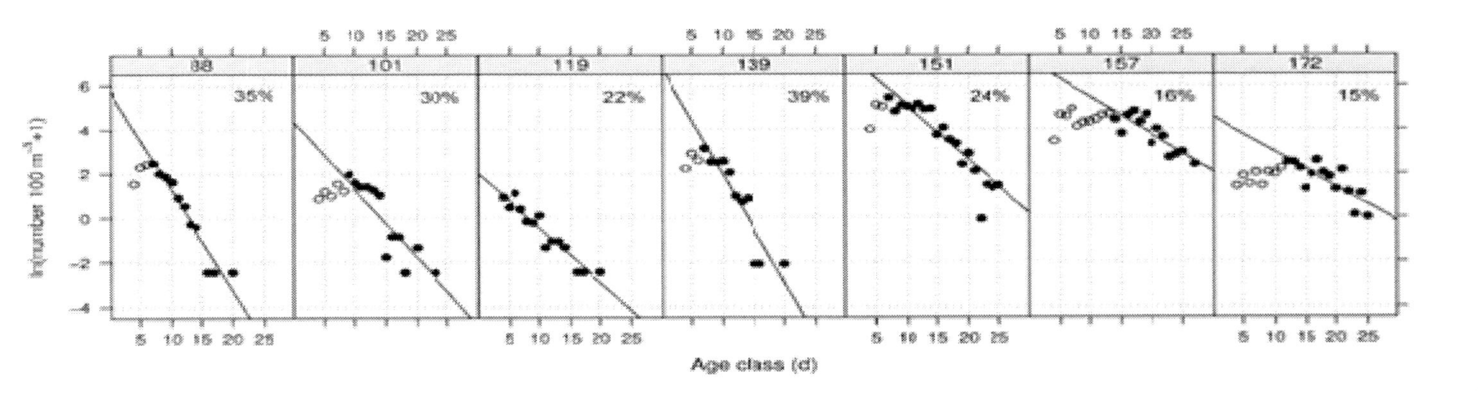

Figure 3. Age-specific mortality curves for Twaite shad larvae in the River Mira (Portugal) during the year 2000 (Esteves [87]). The abundance estimates for each age-class (circles) were derived from age-length keys. Number above each plot is day of sampling (from 1 January) and inside panels is mortality (% d^{-1}). Age classes used to estimate mortality rates are represented by filled circles. Days of year 101 and 151 corresponded to 10 April and 30 May, respectively.

Alosines emigrating seaward actively feed on zooplankton. In the freshwater reaches of the rivers Wye (UK) and Sebou (Morocco), the diet of twaite shad larvae is mostly comprised of larval aquatic insects and accessorily of crustacean zooplankton [74, 80]. Twaite shad larvae in the rivers Wye and Usk (UK) specialized upon chironomid larvae, with chironomid pupae, *Alona* spp. (Cladocera), cyclopoid copepods and ephemeropterans larvae also eaten by many fish in smaller numbers. In River Towy (UK), however, twaite shad larvae clearly preferred cyclopoid copepods, which accounted for 96% of all prey consumed [81]. Hilsa up to 20-40 mm TL feed mostly on diatoms and sparingly on copepods, *Daphnia* spp, and ostracods [7]. Despite the differences specific to each habitat's plankton composition, the more recent studies seem to confirm Aprahamian [69] and Cassou-Leins and Cassou-Leins [74] earlier findings.

Surface temperature, salinity, turbidity and/or rainfall(a proxy of river flow) were important factors in determining the abundance of twaite shad yolk-sac larvae and larvae abundance in the upper River Mira (Portugal). They were included as covariates in the generalized additive models obtained by Esteves and Andrade [70] for twaite shad larvae. Additionally, microplankton biomass and the densities of crustacean' *nauplii*, zooplankters' (isopods, small insects, cladocerans and copepodites) and competitors/predators' (medusae, mysids and amphipods) clearly affected the abundance of larval shad in that river. Thus the distribution of twaite shad larvae observed in that study is partly related to spatial and/or seasonal displacements of specimens as a function of potential prey and clearly avoiding predators. Using canonical correspondence analysis, Gerkens and Thiel [77] found that twaite shad larvae from ElbeRiver(Germany) preferred maximum water depth and distance from the shore line and short distances to the river channel.

Published information on age, growth, mortality or condition of shad larvae other than allis or American shads is anecdotal, consisting mainly of sizes-at-hatch or at-age and qualitative assessments [26, 67, 82]. In contrast, those traits have been investigated in situ on the early life-history stages of closely-related species, e.g. allis shad in the Gironde-Garonne-Dordogne watershed (France) [83] and the American shad in the Connecticut River (USA) [41, 84, 85] and in the Hudson River (USA) [44, 45].

In the Hoogly Estuary (India), growthof hilsa was estimated to be 15-20 mm month^{-1} during the initial 2-3 months after hatching [7]. From the sizes at age referred to by Wang [30] for Reeves shad larvae, I estimated a growth rate of 0.53 mm d^{-1} for the first 5 days after hatching. Recently, Hook et al [86] estimated growth rates for larval alewifes 3.5-30 mm TL from two lakes in the Laurentian Great Lakes region to be 0.85-0.91 mm d^{-1}. These estimates are relatively higher than previously reported. Moreover, those authors calculated that mortality rates were of 0.22-0.30 d^{-1} (i.e. 20-26% d^{-1}). Recently, Lochet et al. [83] validated the daily periodicity of growth increments in the otoliths of hatchery-reared allis shad larvae with an accuracy of 4 days for fish>30 days. Using length-at-age data from their experiment, I estimated that those allis shad larvae grew at rate of 0.27 mm d^{-1}. Esteves [87] found that twaite shad larvae in the River Mira (Portugal) do not grow at the same rate or experience uniform mortality during the spawning season. Date-specific growth rates of shad larvae varied seasonally, from *c.* 0.55 mm d^{-1} in March-April to about 0.33 mm d^{-1} in late-June. On the other hand, date-specific mortality rates of twaite shad larvae declined steadily from 35-39 to 22-15 % d^{-1} during two 30-days intervals, late-March-to-late-April and mid-May-to-late-June (Figure 3). Seasonal changes of and discrepancies among estimates of growth and mortality rates have been attributed to differences in ambient conditions (including different temperatures). Esteves [87]findingsdiffer from results of Crecco and Savoy[41] for American shad in the Connecticut River (USA). Using a different methodology to estimate mortalities, those authors found that cohort-specific mortality rates were higher for early-spawned cohorts (*c.* 6 to 7 % d^{-1}) and declined gradually to about 2-3 % d^{-1} later in the spawning season.

To further understand the ecology of larval shad, Esteves et al. [88] assessed the nutritional condition of twaite larvae in River Mira (Portugal) using individual RNA/DNA ratios. The RNA/DNA ratio is an ecophysiological index of condition that reflects the potential for protein synthesis[89] and is therefore susceptible to changes in the environment. The amount of DNA in a cell is constant; the amount of RNA indicates how actively the cell is synthesising proteins, and thus perhaps how healthy it is [90, 91]. Short-term changes in feeding activity and development rates are reflected in RNA/DNA level after a few hours to within a few days [92-94]. There is also evidence from laboratory and field studies that a decrease in average nucleic acid ratios is associated with reduced survival[95-97] and growth rates [98]. Esteves et al. [88] found that RNA/DNA ratios of 7-20 mm SL larvae of twaite shad varied throughout the day being significantly lower than average during dawn and dusk hours. The likely factors responsible for these results remained unclear. The authors then obtained generalized additive models of seasonal data (February 1998-July 2000) wherein particular environmental conditions, namely water temperatures in the range 21-23 °C and freshwater pulses and/or turbidity (<2 mg DW m^{-3}), largely contribute to create adequate environments, e.g. in terms of potential prey abundance, which likely enhance the nutritional condition of twaite shad larvae during the period they remain in the upper reaches of River Mira.

JUVENILE ALOSINES

At the onset of the juvenile period nearly all organs are present and the fish has the appearance of a small adult. Metamorphosis occurs at about 28 mm TL for American shad [47]. At 20 mm in length, allis shad have already metamorphosed into juveniles[73] whereas twaite shad are still in a "transitional stage" [67]. During juvenile stage growth is the dominant process, wherein a fish can increase its weight by 10^3 to 10^6 times [72]. Houde [99]reasoned that the juvenile stage is more important to recruitment of freshwater fish species than the previous larval stage as opposed to marine fish species. Juveniles of anadromous alosines develop in freshwater or brackish water and migrate to the ocean, where they spend 3 to 7 years before returning to freshwater, as mature adults, to spawn [59, 100].

Juvenile American shad were observed throughout the water column in the Columbia River (USA) using trawl and hydroacoustic surveys [39]. The results from their surveys suggest that juvenile American shad stayed near the bottom or were in inshore areas during the day and dispersed throughout the water column at night. American shad juveniles demonstrate temporal and latitudinal migration trends; nonetheless they seem move downstream of the spawning grounds during fall, particularly at night (18:00-23:00) [53].Gahagan et al. [101] found that most of the juveniles of alewife in Bride Lake (USA) that migrated seaward from mid-June to mid-August departed in three 1- or 2-d pulses. Moreover, the diel distribution of migration had two peaks: around dawn (05:00-07:00) and, to a lesser extent, in mid-day (11:00-13:00). Blueback herring migrate to seawater in a single period late-September-October coincident with sharp autumnal decrease in temperature. On the other hand, the peaks in alewife migration – early and late-season migratory pulses – reported by Iafrate and Oliveira [51] should mitigate competition and low food availability or winter kill due to dropping water temperatures. Age-0 allis shad migrate seaward during the summer and fall of their first year of life [9]. In the Severn and Elbe rivers, the majority of twaite shad leave the estuary by the end of October, while in the River Garonne it is not until the end of February that the majority have migrated seaward [26]. Most YOY of the River Rhône subspecies of twaite shad *A. f. rhodanensis* occurred in the Vaccarés Lagoon (one compartment of the Rhône delta aquatic system closer to marine areas) from mid-June to mid-July and rarely in August or September [102]. The seaward migration of allis and twaite shads seems to occur earlier in southern rivers. The residence time in the estuary is not well known [103]. Recently, from otolith Sr:Ca ratios and daily growth increments, Lochet et al. [104] retrospectively estimate that allis shad entered the estuary of an older age than twaite shad (88 d. v. 59 d), stayed shorter in the estuary (11 d. v. 17 d.), and exited to the sea at an older age (99 d. v. 77 d.).

Juvenile migration is modulated by water temperature, river discharge and biological factors such as size and level of adaptability to marine conditions. In American shad, the mechanisms of initiating downstream migration are not yet clearly understood. O'Donnell and Letcher [48] list the hypothesized, possible cues for migration: flow, lunar cycle, water temperature, size, osmoregulatory physiology and age. Since early-hatched American shad in Connecticut River (USA) were missing from samples of outmigrating juveniles, O'Donnell and Letcher[48] proposed that this pattern is consistent with the hypothesis that fish move downstream as they grow and age.In addition, temperature and previous day's rainfall, date and discharge were the best predictors of migration rate of alewife juveniles in Bride Lake (USA) [101]. Boisneau et al. [36] found that climatic change upon the thermal and hydrological regimes of River Loire (France) between 1976 and 1995 antecipated in 17 d the start of the seaward migration of YOY allis shad.

Due to the proximity of the estuary areas and forced spawning sites, and the important drift of eggs and larvae[76], young allis shad may reach brackish waters very early in their ontogeny [105].Therein, high or complete mortality is expected to be induced by direct seawater exposure of"unprepared" individuals[105-107]. Seawater tolerance of shads is developed during the larval-juvenile transition: 18-70 d in allis shad [108]; 36-45 dph in American shad [107]. In contrast to American shad, which lose the ability to osmoregulate during juvenile stage [109], allis shad have the continued ability to hypoosmoregulate and thus adapt to hyperosmotic environments and withstand freshwater for longer periods [108].

Before emigrating to the ocean, juveniles of anadromous alosines develop in freshwater or brackish water for a few months. During that period, the diets of 0+ juveniles of twaite shad in the River Wye (UK) were dominated by chironomid larvae, with chironomid pupae, ephemeropterans larvae, adult dipterans and simulid larvae also consumed by many individuals. In addition, 18% of the juvenile had preyed upon fish, consuming chub *Leuciscus cephalus*, minnow *Phoxinus phoxinus* and bleak *Alburnus alburnus*[81]. Oesmann and Thiel [110] identified copepods, mysids and fish as the main food item in the stomachs of juvenile twaite shad individuals in the Elbe Estuary (Germany). These results are in line with those of Grabe [111] for juvenile alewifes, blueback herring and American shad at inshore location(s) on the lower Hudson River (USA): alewifes feed primarily on chironomids and the amphipod *Corophium lacustre*, shad on formicids and larval chironomid, and herring on chironmids and copepods. The diet of American shad in the John Day Reservoir consisted mainly of cyclopoid copepods, and to much lesser extent included calanoid copepods, cladocerans, and dipteran larvae [39]. Grabe [111] found that diel variation in prey selection was evident; blueback herring feeding less intensely at night and shad feeding less at sunrise when compared to other periods of the day (sunset, day, and night). Juvenile hilsa up to 100 mm TL feed mainly on small crustacean (copepods, *Daphnia* spp., and ostracods), insects larvae, chironomid larvae and Polyzoa, whilst bigger specimens (up to 150 mm TL) included small shrimps in their diet [7]. As with larvae, despite the differences pertaining to habitat, juvenile shad consume insect larvae and crustacean zooplankton namely copepods; fishes are now included.

A relatively long list of parasites infect 0+ twaite shad in UK rivers [81]: *Apiosoma* sp. (Protozoa) dominated (was observed in over 60% of estuarine 0+ juveniles) but *Gyrodactylus* sp. (Monogenea), Proleptinae (Nematoda) larvae, *Pomphorhynchus laevis* (Acanthocephala), *Spinitectus* sp. (Nematoda) larvae and *Trichodina* sp. (Protozoa) were also observed. The level of infection did not affect juveniles since there was no difference in condition between parasitized and non-paratisized 0+ twaite shad.

Growth is the dominant process during the juvenile stage of fishes [72]. In the Hoogly Estuary (India), growth of hilsa was estimated to be 10 mm month^{-1}after the initial 2-3 months following hatching [7].In Bride Lake (USA), the length at mean age of migrant alewife was 11% greater than that of nonmigrant specimens, despite the non-significant difference in growth rates, estimated at 0.294 mm d^{-1}[101]. The growth rate of YOY of the River Rhône subspecies of twaite shad *A. f. rhodanensis* varied between 0.9 and 1.4 mm d^{-1}. Neither the temperature (in degree-days) nor the salinity, nor the density of YOY shad (expressed as CPUE) could explain the differences among years studied [102]. Using length-at-age data extracted from Figure 5 in Aprahamian et al. [26] for four rivers, Severn (UK), Elbe (Germany), Gironde (France), and Sebou (Morocco), I estimated (linear) growth rates for juvenile twaite shad that ranged from 2.93 to 12.79 mm month^{-1}, i.e. 0.10 mm d^{-1} to 0.43 mm d^{-1} if month=30 days (Figure 4). Alternatively, the Gompertz and logarithmic equations were fit to data from the River Elbe and River Sebou, respectively: $y = 88.31 \cdot \exp(-5.20 \cdot \exp(-0.33x))$ and $y = 27.84\ln(x) + 11.74$ (where y is length (mm) and x is age (month)). These statistically significant models (F-test, p=0.0142 and p=0.0013, respectively) also appropriately described the size increment of twaite shad but are more complex and not readily comparable thus the initial approach using linear regression. Using a standard bioenergetics model parameterized for juvenile twaite shad (0.5 to 1.5 g in body weight), and assuming consumption rate is constrained by dissolved oxygen, Maes et al. [35] predicted increased growth rate potential of YOY in the lower estuary and nearby the historical spawning sites upstream in River Scheldt (Belgium, Netherlands). There is a reasonable possibility that this particular population of twaite shad recovers by 2010 due to the expected environmental recovery. Wang [30] presents a linear model, $y = 32.25 + 0.59x$, to describe the relationship between body length (y, mm) of Reeves shad juveniles from the PuoyangLake (China) and time (x, days). Its slope can be interpreted as the growth rate, i.e. 0.59 mm d^{-1}.

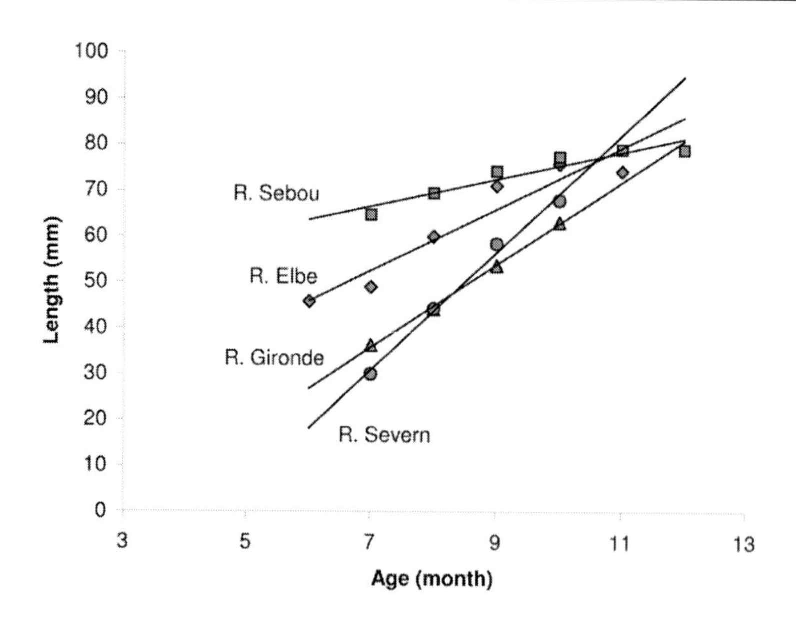

Figure 4. Length of age-0 twaite shad from four rivers, R. Sebou (Morocco) y=2.93x+46.09 (R^2=0.903; p=0.0036), R. Elbe (Germany) y=6.68x+5.88 (R^2=0.911; p=0.0030), R. Gironde (France) y=9.00x-27.16 (R^2=0.998; p=0.0009) and R. Severn (UK) y=12.79x-58.58 (R^2=0.992; 0.0041) (see main text for details; data was extracted from Aprahamian et al. [26]).

Conclusion

There is much more work done on the early life-history stages of North American species, both in the laboratory and in situ, particularly on American shad and alewife, than studies dealing with European species. In this chapter, I compiled information published about eggs, larvae and juveniles after the two international conferences on alosines held in France and USA a decade ago. Since then a number of topics have been addressed: changes in diel and seasonal distribution and nutritional condition of twaite shad larvae and their relationships with environmental variables were statistically modeled; positive photoresponse was observed, daily periodicity of increment deposition on otoliths was validated and salinity tolerance was studied experimentally in allis shad larvae and juveniles; diet compositionand parasites of larval twaite shad were assessed; duration of the estuarine phase and emigrating patterns were studied for allis and twaite shad, river herringand American shad; tools for the discrimination of natal origin of allis shad and American shadwere developed; no effect of estuarine residence time of juveniles on mercury contamination of adult allis shad was found; and climatic factors, through their effect upon the hydrological and thermal regimes of rivers, were found to influence the seaward migration of YOY allis shad but not recruitment.

Despite several recent contributions to augment knowledge of anadromous shads' early life-history stages, particularly of European species, work is still necessary to address yet unanswered questions dealing with: fundamental aspects of biology, traits such as growth and mortality, in less-studied species, namely of Mediterranean and Ponto-caspian species, Indian shads and hilsa; the role of larval and juvenile shad in the food web; and (following Limburg etal [28]) metrics describing the welfare of YOY (abundance, size, condition, location) that could readily serve as a component ofa sustainability barometer in water shed management.

REFERENCES

[1] Alexandrino, P. (1995). Genetic and morphological differentiation among some Portuguese populations of allis shad *Alosa alosa* (L., 1758) and twaite shad *Alosa fallax* (Lacépède, 1803). *Publicaciones Especiales del Instituto Español de Oceanografía, 21*, 15-24.

[2] Mennesson-Boisneau, C., Aprahamian, C. D., Sabatié, M. R. and Cassou-Leins, J. J. (2000). Caractéristiques des adultes. In J.-L. Baglinière and P. Elie (Eds.). *Les aloses (Alosa alosa et Alosa fallax spp.)* (pp. 33-53) Paris, France: INRA Editions et Cemagref Editions.

[3] Baglinière, J.-L. (2000). Le genre*Alosa* sp. In J.-L. Baglinière and P. Elie (Eds.). *Les aloses (Alosa alosa et Alosa fallax spp.)* (pp. 3-30) Paris, France: INRA Editions et Cemagref Editions.

[4] Waldman, J. R. (2003). Introduction to the shads. *American Fisheries Society Symposium, 35*, 3-9.

[5] Hendricks, M. L., Hoopes, R. L., Arnold, D. A. and Kaufman, M. L. (2002). Homing of hatchery-reared American shad to the Lehigh River, a tributary to the Delaware River. *North American Journal of Fisheries Management, 22*, 243-248.

[6] Tomás, J., Augagneur, S. and Rochard, E. (2005). Discrimination of the natal origin of young-of-the-year Allis shad (*Alosa alosa*) in the Garonne-Dordogne basin (south-west France) using otolith chemistry. *Ecology of Freshwater Fish, 14*, 185-190.

[7] Rahman, M. J. (2006). Recent advances in the biology and management of Indian shad (*Tenualosa ilisha* Ham.). *SAARC Journal of Agriculture, 4*, 76-98.

[8] Limburg, K. E. (2001). Through the gauntlet again: demographic restructuring of American shad by migration. *Ecology, 82*, 1584-1596.

[9] Baglinière, J.-L., Sabatié, M. R., Rochard, E., Alexandrino, P. and Aprahamian, M. W. (2003). The Allis shad *Alosa alosa*: biology, ecology, range, and status of populations. *American Fisheries Society Symposium, 35*, 85-102.

[10] Dufour, E., Hook, T. O., Patterson, W. P. and Rutherford, E. S. (2008). High-resolution isotope analysis of young alewife *Alosa pseudoharengus* otoliths: assessment of temporal resolution and reconstruction of habitat occupancy and thermal history. *Journal of Fish Biology, 73*, 2434-2451.

[11] Blaber, S. J. M. (2003). Biology, fisheries, and status of tropical shads *Tenualosa* spp. in South and Southeast Asia. *American Fisheries Society Symposium, 35*, 49-58.

[12] Alexandrino, P., Faria, R., Linhares, D., Le Corre, M., Sabatié, R., Baglinière, J. L. and Weiss, S. (2006). Interspecific differentiation and intraspecific substructure in two closely related clupeids with extensive hybridization, *Alosa alosa* and *Alosa fallax*. *Journal of Fish Biology, 69 (Suppl. B)*, 242-259.

[13] Walther, B. D. and Thorrold, S. R. (2008). Continental-scale variation in otolith geochemistry of juvenile America shad (*Alosa sapidissima*). *Canadian Journal of Fisheries and Aquatic Sciences, 65*, 2623-2635.

[14] Waters, J. M., Epifanio, J. M., Gunter, T. and Brown, B. L. (2000). Homing behaviour facilitates subtle genetic differentiation among river populations of *Alosa sapidissima*: microsatellites and mtDNA. *Journal of Fish Biology*, *56*, 622-636.

[15] McDowall, R. M. (2003). Shads and diadromy: implications for ecology, evolution, and biogeography. *American Fisheries Society Symposium*, *35*, 11-23.

[16] Milton, D. A. and Chenery, S. R. (2003). Movement patterns of the tropical shad hilsa (*Tenualosa ilisha*) inferred from transects of 87Sr/86Sr isotope ratios in their otoliths. *Canadian Journal of Fisheries and Aquatic Sciences, 60*, 1376-1385.

[17] Dodson, J. J. and Leggett, W. C. (1974). Role of olfaction and vision in the behavior of American shad (*Alosa sapidissima*) homing to the Connecticut River from Long Island Sound. *Journal of the Fisheries Research Board of Canada, 31*, 1607-1619.

[18] Assis, C. A. (1990). Threats to the survival of anadromous fishes in the River Tagus, Portugal. *Journal of Fish Biology, 37 (Suppl. A)*, 225-226.

[19] Crecco, V. and Savoy, T. (1984). Effects of fluctuations in hydrographic conditions on year-class strength of American shad (*Alosa sapidissima*) in the Connecticut river. *Canadian Journal of Fisheries and Aquatic Sciences*, *41*, 1216-1233.

[20] Taverny, C., Belaud, A., Elie, P. and Sabatié, M. R. (2000). Influence des activités humaines. In J.-L. Baglinière and P. Elie (Eds.). *Les aloses (Alosa alosa et Alosa fallax spp.)* (pp. 227-248) Paris, France: INRA Editions et Cemagref Editions.

[21] Limburg, K. E. and Waldman, J. R. (2009). Dramatic declines in North Atlantic diadromous fishes. *BioScience*, *59*, 955-965.

[22] Novadoru, I. and Waldman, J. R. (2003). Shads of Eastern Europe from the Black Sea: Review of species and fisheries. *American Fisheries Society Symposium*, *35*, 69-76.

[23] Alexandrino, P. (1996). Estudo de populações de sável (*Alosa alosa* L.) e savelha (*Alosa fallax* Lacépède). Análise da diferenciação interespecífica, subestruturação e hibridação. (Doctoral Thesis) Faculdade de Ciências, Universidade do Porto, Porto, Portugal.185 p.

[24] Larinier, M., Travade, F. and Dartiguelongue, J. (2000). La conception des dispositifs de franchissement. In J.-L. Baglinière and P. Elie (Eds.). *Les aloses (Alosa alosa et Alosa fallax spp.)* (pp. 249-259) Paris, France: INRA Editions et Cemagref Editions.

[25] Hendricks, M. L., Hoopes, R. L., Arnold, D. A. and Kaufman, M. L. (2003). Culture and transplant of alosines in North America. *American Fisheries Society Symposium*, *35*, 303-312.

[26] Aprahamian, M. W., Baglinière, J.-L., Sabatié, M. R., Alexandrino, P., Thiel, R. and Aprahamian, C. D. (2003). Biology, status, and conservation of the anadromous Atlantic Twaite shad *Alosa fallax fallax*. *American Fisheries Society Symposium*, *35*, 103-124.

[27] Olney, J. E. and Hoenig, J. M. (2001). Managing a fishery under Moratorium: Assessment opportunities for Virginia's stocks of American shad. *Fisheries*, *26*, 6-12.

[28] Limburg, K. E., Hattala, K. A. and Kahnle, A. (2003). American shad in its native range. *American Fisheries Society Symposium*, *35*, 125-140.

[29] Schmidt, R. E., Jessop, B. M. and Hightower, J. E. (2003). Status of river herring stocks in large rivers. *American Fisheries Society Symposium*, *35*, 171-182.

[30] Wang, H.-P. (2003). Biology, population dynamics, and culture of Reeves shad *Tenualosa reevesii*. *American Fisheries Society Symposium*, *35*, 77-83.

[31] CIFRI (2007). Hoogly Estuary. India, Central Inland Fisheries Research Insitute (leaflet) online at http://www.cifri.ernet.in/17.pdf (accessed on February 17, 2011).

[32] Legall, O. (2000). Origine et histoire des aloses. In J.-L. Baglinière and P. Elie (Eds.). *Les aloses (Alosa alosa et Alosa fallax spp.)* (pp. 127-136) Paris, France: INRA Editions et Cemagref Editions.

[33] Alexandrino, P. and Boisneau, P. (2000). Diversité génétique. In J.-L. Baglinière and P. Elie (Eds.). *Les aloses (Alosa alosa et Alosa fallax spp.)* (pp. 179-196) Paris, France: INRA Editions et Cemagref Editions.

[34] Sabatié, M. R., Boisneau, P. and Alexandrino, P. (2000). Variabilité morphologique. In J.-L. Baglinière and P. Elie (Eds.). *Les aloses (Alosa alosa et Alosa fallax spp.)* (pp. 137-178) Paris, France: INRA Editions et Cemagref Editions.

[35] Maes, J., Stevens, M. and Breine, J. (2008). Poor water quality constrains the distribution and movements of twaite shad *Alosa fallax fallax* (Lacépède, 1803) in the watershed of river Scheldt. *Hydrobiologia*, *602*, 129-143.

[36] Boisneau, C., Moatar, F., Bodin, M. and Boisneau, P. (2008). Does global waming impact on migration patterns and recruitment of Allis shad (*Alosa alosa* L.) young of the year in the Loire River, France? *Hydrobiologia*, *602*, 179-186.

[37] Baglinière, J.-L., Sabatié, M. R., Alexandrino, P., Aprahamian, M. W. and Elie, P. (2000). Les aloses: une richesse patrimoniale à conserver et à valoriser. In J.-L. Baglinière and P. Elie (Eds.). *Les aloses (Alosa alosa et Alosa fallax spp.)* (pp. 263-275) Paris, France: INRA Editions et Cemagref Editions.

[38] Coad, B. W., Hussain, N. A., Ali, T. S. andLimburg, K. E. (2003). Middle Eastern shads. *American Fisheries Society Symposium*, *35*, 59-67.

[39] Petersen, J. H., Hinrichsen, R. A., Gadomski, D. M., Feil, D. H. and Rondorf, D. W. (2003). American shad in the Columbia River. *American Fisheries Society Symposium*, *35*, 141-155.

[40] Wang, H.-P., Xiong, B.-X., Wei, K.-J. andYao, H. (2003). Broodstock Rearing and Controlled Reproduction of Reeves Shad *Tenualosa reevesii*. *Journal of the World Aquaculture Society*, *34*, 308-318.

[41] Crecco, V. and Savoy, T. (1987). Effects of climatic and density-dependent factors on intra-annual mortality of larval American shad. *American Fisheries Society Symposium*, *2*, 69-81.

[42] Crecco, V. and Blake, M. M. (1983). Feeding ecology of coexisting larvae of American shad and blueback herring in the Connecticut River. *Transactions of the American Fisheries Society*, *112*, 498-507.

[43] Crecco, V., Savoy, T. and Whitworth, W. (1986). Effects of density-dependent and climatic factors on American shad, *Alosa sapidissima*, recruitment: a predictive approach. *Canadian Journal of Fisheries and Aquatic Sciences*, *43*, 457-463.

[44] Limburg, K. E. (1996). Growth and migration of 0-year American shad (*Alosa sapidissima*) in the Hudson River estuary: otolith microstructural analysis. *Canadian Journal of Fisheries and Aquatic Sciences*, *53*, 220-238.

[45] Limburg, K. E. (1996). Modelling the ecological constraints on growth and movement of juvenile American shad (*Alosa sapidissima*) in the Hudson River Estuary. *Estuaries*, *19*, 794-813.

[46] Savoy, T. and Crecco, V. (1987). Daily increments on the otoliths of larval American shad and their potential use in population dynamics studies. In R. C. Summerfelt and G.

E. Hall (Eds.). *The age and growth of fish* (pp. 413-431) Ames, Iowa, USA: The Iowa State University Press.

[47] Savoy, T. and Crecco, V. (1988). The timing and significance of density-dependent and density-independent mortality of American shad, *Alosa sapidissima. Fishery Bulletin, 86*, 467-481.

[48] O'Donnell, M. J. and Letcher, B. H. (2008). Size and age distributions of juvenile Connecticut River American shad above HaddleyFalls: influence on outmigratiion representation and timing. *River Research and Applications, 24*, 929-940.

[49] Hoffman, J. C., Limburg, K. E., Bronk, D. A. and Olney, J. E. (2008). Overwintering habitats of migratory juvenile American shad in Chesapeake Bay. *Hydrobiologia, 81*, 329-345.

[50] Mickle, P. F., Schaefer, J. F., Adams, S. B. and Kreiser, B. R. (2010). Habitat use of age 0 Alabama shad in the PascagoulaRiver drainage, USA. *Ecology of Freshwater Fish, 19*, 107-115.

[51] Iafrate, J. and Oliveira, K. (2008). Factors affecting migration patterns of juvenile river herring in a coastal Massachusetts stream. *Environmental Biology of Fish, 81*, 101-110.

[52] IUCN (2008). *IUCN red list categories.*Gland, Switzerland, International Union for Conservation of Nature, Species Survival Commission (online at http://www.iucnredlist.org/, accessed February 17, 2011)

[53] Greene, K. E., Zimmerman, J. L., Laney, R. W. and Thomas-Blate, J. C. (2009). AtlanticCoast Diadromous Fish Habitat:A Review of Utilization, Threats, Recommendations for Conservation, and Research Needs. Atlantic States Marine Fisheries Commission Habitat Management Series No. 9. Washington, D.C.464 p.

[54] Limburg, K. E., Gibson, J., Pine, B., Quinn, T. and Sands, N. J. (2007). Terms of Reference and Advisory Report to the American Shad Stock Assessment Peer Review. Atlantic States Marine Fisheries Commission. Stock Assessment Report No. 07-01.

[55] Baglinière, J. L., Rochard, E. and Vigneux, E. (2001). First international conference on European shads. *Bulletin Français de la Pêche et de la Pisciculture, 362/363*, 384 p.

[56] Limburg, K. E. and Waldman, J. R. (2003). *Biodiversity, status, and conservation of the world's shads*. Bethesda, Maryland, USA: American Fisheries Society.

[57] Olney, J. E. and McBride, R. S. (2003). Intraspecific variation in batch fecundity of American shad: revisiting the paradigm of reciprocal latitudinal trends in reproductive traits. *American Fisheries Society Symposium, 35*, 185-192.

[58] Leggett, W. C. and Carscaden, J. E. (1978). Latitudinal variation in reproductive characteristics of American shad (*Alosa sapidissima*): evidence for population specific life history strategies in fish. *Journal of the Fisheries Research Board of Canada, 35*, 1469-1478.

[59] Mennesson-Boisneau, C., Aprahamian, C. D., Sabatié, M. R. and Cassou-Leins, J. J. (2000). Remontée migratoire des adultes. In J.-L. Baglinière and P. Elie (Eds.). *Les aloses (Alosa alosa et Alosa fallax spp.)* (pp. 55-72) Paris, France: INRA Editions et Cemagref Editions.

[60] Aprahamian, M. W. (1981*). Aspects of the biology of the twaite shad (Alosa fallax) in the rivers Severn and Wye*. Proceedings of the 2nd British Freshwater Fisheries Conference, University of Liverpool, pp. 111-119.

[61] Mettee, M. F. and O'Neil, P. E. (2003). Status of Alabama shad and skipjack herring in Gulf of Mexico drainages. *American Fisheries Society Symposium, 35*, 157-170.

[62] Thiel, R., Sepúlveda, A. and Oesmann, S. (1996). Occurrence and distribution of twaite shad (*Alosa fallax* Lacepède) in the lower Elbe River, Germany. In A. Kirchhoffer and D. Hefti (Eds.). *Conservation of endangered freshwater fishes in Europe* (pp. 157-170) Basel, Switzerland: Birkhauser Verlag.

[63] Taverny, C. (1991). Contribution à la connaissance de la dynamique des populations d'aloses (*Alosa alosa* et *Alosa fallax*) dans le système fluvio-estuarien de la Gironde: Pêche, biologie et écologie. Étude particulière de la dévailason et de l'impact des activités humaines. (Doctoral Thesis) Université de Bordeaux I, Bordeaux.375 p.

[64] Caswell, P. A. and Aprahamian, M. W. (2001). Use of river habitat survey to determine the spawning habitat characteristics of twaite shad (*Alosa fallax fallax*). *Bulletin Français de la Pêche et de la Pisciculture, 362/363*, 919-929.

[65] Cassou-Leins, J.-J., Cassou-Leins, F., Boisneau, P. and Baglinière, J.-L. (2000). La reproduction. In J.-L. Baglinière and P. Elie (Eds.). *Les aloses (Alosa alosa et Alosa fallax spp.)* (pp. 73-92) Paris, France: Cemagref Editions et INRA Editions.

[66] Aprahamian, M. W. (1982). Aspects of the biology of the twaite shad, *Alosa fallax fallax* (Lacépède), in the Rivers Severn and Wye. (Doctoral Thesis) University of Liverpool, Liverpool, UK.

[67] Quignard, J. P. and Douchement, C. (1991). *Alosa fallax fallax* (Lacepède, 1803). In H. Hoestland (Eds.). *The freshwater fishes of Europe. Vol. 2. Clupeidae, Anguillidae* (pp. 225-253), Wiesbaden: AULA-VerlagWiesbaden.

[68] Ross, R. M., Bennett, R. M. and Backman, T. W. H. (1993). Habitat use by spawning adult, egg, and larval American shad in the Delaware river. *Rivers, 4*, 227-238.

[69] Aprahamian, M. W. (1989). The diet of juvenile and adult twaite shad Alosa fallax fallax (lacépède) from the rivers Severn and Wye (Britain). *Hydrobiologia, 179*, 173-182.

[70] Esteves, E. and Andrade, J. P. (2008). Diel and seasonal distribution patterns of eggs, embryos and larvae of Twaite shad *Alosa fallax fallax* (Lacépède, 1803) in a lowland tidal river. *Acta Oecologica, 34*, 172-185.

[71] Novadoru, I. (2001). Seaward drift og the pontic shad larvae (*Alosa pontica*) and the influence of Danube river hydrology on their travel path through the Danube delta system. *Bulletin Français de la Pêche et de la Pisciculture, 362/363*, 749-760.

[72] Fuiman, L. A. (2002). Special considerations of fish eggs and larvae. In L. A. Fuiman and R. G. Werner (Eds.). *Fishery Science. The unique contributions of early life stages* (pp. 1-32) Oxford, UK: Blackwell Science/Publishing.

[73] Taverny, C., Cassou-Leins, J. J., Cassou-Leins, F. and Elie, P. (2000). De l'ouef à l'adulte en mer. In J.-L. Baglinière and P. Elie (Eds.). *Les aloses (Alosa alosa et Alosa fallax spp.)* (pp. 93-124) Paris, France: INRA Editions et Cemagref Editions.

[74] Cassou-Leins, F. and Cassou-Leins, J.-J. (1981). Recherches sur la biologie et l'halieutique des migrateurs de la Garonne et principalement de l'alose *Alosa alosa* L., (Doctoral Thesis) Institut National Polytechnique de Toulose, Toulouse.362 p.

[75] Jatteau, P. and Bardonnet, A. (2008). Photoresponse in allis shad larvae. *Journal of Fish Biology, 72*, 742-746.

[76] Véron, V., Jatteau, P. and Bardonnet, A. (2003). First results on the behaviour of young stages of allis shad *Alosa alosa. American Fisheries Society Symposium, 35*, 241-251.

[77] Gerkens, M. and Thiel, R. (2001). Habitat use of age-0 twaite shad (*Alosa fallax* Lacépède, 1803) in the tidal freshwater region of the Elbe River, Germany. *Bulletin Français de la Pêche et de la Pisciculture, 362/363*, 773-784.

[78] Gadomski, D. M. and Barfoot, C. A. (1998). Diel and distributional abundance patterns of fish embryos and larvae in the lower Columbia and Deschutes rivers. *Environmental Biology of Fishes, 51*, 353-368.

[79] BOBP (1987). *Hilsa investigations in Bangladesh*. Colombo, Sri Lanka, Bay of Bengal Programme (BOBP/REP/36), Marine Fishery Resources Management (RAS/81/051), FAO-UNDP.

[80] Aprahamian, M. W. (1989). The diet of juvenile and adult twaite shad *Alosa fallax fallax* (Lacépède) from the Rivers Severn and Wye (Britain). *Hydrobiologia, 179*, 173-182.

[81] Nunn, A. D., Noble, R. A. A., Harvey, J. P. and Cowx, I. G. (2008). The diets and parasites of larval and 0+ juvenile twaite shad in the lower reaches and estuaries of the rivers Wye, Usk and Towy, UK. *Hydrobiologia, 614*, 209-218.

[82] Baglinière, J.-L. and Elie, P. (2000). *Les aloses (Alosa alosa et Alosa fallax spp.)*. Paris, France: Cemagref Editions et INRA Editions.

[83] Lochet, A., Jatteau, P., Tomás, J. and Rochard, E. (2008). Retrospective approach to investigating the early life history of a diadromous fish: allis shad *Alosa alosa* (L.) in the Gironde–Garonne–Dordogne watershed. *Journal of Fish Biology, 72*, 946-960.

[84] Crecco, V. and Savoy, T. (1985). Effects of biotic and abiotic factors on growth and relative survival of young American shad, *Alosa sapidissima*, in the Connecticut River. *Canadian Journal of Fisheries and Aquatic Sciences, 42*, 1640-1648.

[85] Crecco, V. and Savoy, T. (1987). Review of recruitment mechanisms of the American shad: the critical period and match-mismatch hypothesis reexamined. *American Fisheries Society Symposium, 1*, 455-468.

[86] Hook, T. O., Rutherford, E. S., Maison, D. M. and Carter, G. S. (2007). Hatch dates, growth, survival, and overwinter mortality of age-0 alewifes in Lake Michigan: implications for habitat-specific recruitment success. *Transactions of the American Fisheries Society, 136*, 1298-1312.

[87] Esteves, E. (2006). Ecology of the early life-history stages of shad *Alosa fallax fallax* (Lacépède, 1803) in the River Mira, with a note on *Alosa* spp. larvae in the River Guadiana. (Doctoral Thesis) Faculdade de Ciências do Mar e Ambiente, Universidade do Algarve, Faro.157 p.

[88] Esteves, E., Pina, T. and Andrade, J. P. (2009). Diel and seasonal changes in nutritional condition of the anadromous Twaite shad *Alosa fallax fallax* (Lacépède, 1803) larvae*Ecology of Freshwater Fish, 18*, 132-144.

[89] Chícharo, M. A. (1997). Starvation percentages in field caught *Sardina pilchardus* larvae off southern Portugal. *Scientia Marina, 61*, 507-516.

[90] Clemmesen, C. (1994). The effect of food availability, age or size on the RNA/DNA ratio of individually measured herring larvae: laboratory calibration. *Marine Biology, 118*, 377-382.

[91] Buckley, L. J., Turner, S. I., Halavik, T. A., Smigielski, A. S., Drew, S. M. and Laurence, G. C. (1984). Effects of temperature and food availability on growth, survival, and RNA-DNA ratio of larval sand lance (*Ammodytes americanus*). *Marine Ecology Progress Series, 15*, 91-97.

[92] Bergeron, J.-P. (1997). Nucleic acids in ichthyoplankton ecology: a review, with emphasis on recent advances for new perspectives. *Journal of Fish Biology, 51 (Suppl. A)*, 284-302.

[93] Ferron, A. and Leggett, W. C. (1994). An appraisal of condition measures for marine fish larvae. *Advances in Marine Biology, 30*, 217-303.

[94] Clemmesen, C. (1996). Importance and limits of RNA/DNA ratios as a measure of nutritional condition in fish larvae. In Y. Watanabe, Y. Yamashita and Y. Ooseki (Eds.). *Survival strategies in early life stages of marine resources* (pp. 67-82) Roterdam, Holland: A.A. Balkema.

[95] Canino, M. F., Bailey, K. M. and Incze, L. S. (1991). Temporal and geogrpahic differences in feeding and nutritional condition of walleye pollock larvae*Theragra chalcogramma* in Shelikof Strait, Gulf of Alaska. *Marine Ecology Progress Series, 79*, 27-35.

[96] Canino, M. F. (1994). Effects of temperature and food availability on growthand RNA/DNA ratios of walleye pollock *Theragra chalcogramma* (Pallas) eggs and larvae. *Journal of Experimental Marine Biology and Ecology, 175*, 1-16.

[97] Clemmesen, C., Sanchez, R. and Wongtschowski, C. (1997). A regional comparison of the nutritional condition of SW Atlantic anchovy larvae, *Engraulis anchoita*, based on RNA/NDA ratios. *Archives of Fisheries and Marine Research, 45*, 17-43.

[98] Hovenkamp, F. (1990). Growth differences in larval plaice *Pleuronectes platessa* in the Southern Bight of the North Sea as indicated by otolith increments and RNA/DNA ratios. *Marine Ecology Progress Series, 58*, 205-215.

[99] Houde, E. D. (1994). Differences between marine and freshwater fish larvae: implications for recruitment. *ICES Journal of Marine Science, 51*, 91-97.

[100] Maki, K. L., Hoenig, J. M. and Olney, J. E. (2001). Estimating Proportion Mature at Age When Immature Fish are Unavailable for Study, with Application to American Shad in the York River, Virginia. *North American Journal of Fisheries Management, 21*, 703-716.

[101] Gahagan, B. I., Gherard, K. E. and Schultz, E. T. (2010). Environmental and endogenous factors influencing emigration in juvenile anadromous alewives. *Transactions of the American Fisheries Society, 139*, 1069-1082.

[102] Crivelli, A. J. and Poizat, G. (2001). Timing of migration and exceptional growth of YOY *Alosa fallax rhodanensis* (Roule, 1924) in a lagoon in Southern France. *Bulletin Français de la Pêche et de la Pisciculture, 362/363*, 761-772.

[103] Elie, P., Taverny, C., Mennesson-Boisneau, C. and Sabatié, M. R. (2000). L'exploitation halieutique. In J.-L. Baglinière and P. Elie (Eds.). *Les aloses (Alosa alosa et Alosa fallax spp.)* (pp. 199-226) Paris, France: INRA Editions et Cemagref Editions.

[104] Lochet, A., Boutry, S. and Rochard, E. (2009). Estuarine phase during seaward migration for allis shad *Alosa alosa* and twaite shad *Alosa fallax* future spawners. *Ecology of Freshwater Fish, 18,* 323-335.

[105] Bardonnet, A. and Jatteau, P. (2008). Salinity tolerance in young Allis shad larvae (*Alosa alosa* L.). *Ecology of Freshwater Fish, 17*, 193-197.

[106] Leguen, I., Veron, V., Sevellec, C., Azam, D., Sabatie, R., Prunet, P. and Bagliniere, J. L. (2007). Development of hypoosmoregulatory ability in allis shad Alosa alosa. *Journal of Fish Biology, 70*, 630-637.

[107] Zydlewski, J. and McCormick, S. D. (1997). The ontogeny of salinity tolerance in the American shad, *Alosa sapidissima. Canadian Journal of Fisheries and Aquatic Sciences*, *54*, 182-189.

[108] Leguen, I., Véron, V., Sevellec, C., Azam, D., Sabatié, R., Prunet, P. and Baglinière, J. L. (2007). Development of hypoosmoregulatory ability in allis shad Alosa alosa. *Journal of Fish Biology*, *70*, 630-637.

[109] Zydlewski, J. and McCormick, S. D. (1997). The loss of hyperosmoregulatory ability in migrating juvenile American shad, *Alosa sapidissima. Canadian Journal of Fisheries and Aquatic Sciences*, *54*, 2377-2387.

[110] Oesmann, S. and Thiel, R. (2001). Feeding of juvenile twaite shad (*Alosa fallax* Lacépède, 1803) in the Elbe Estuary. *Bulletin Français de la Pêche et de la Pisciculture*, *362/363*, 785-800.

[111] Grabe, S. A. (1996). Feeding chronology and habits of *Alosa* spp. (Clupeidae) juveniles from the lower Hudson Rover estuary, New York. *Environmental Biology of Fish*, *47*, 321-326.

In: Fish Ecology
Editor: Sean P. Dempsey
ISBN 978-1-61324-282-7

Chapter 9

HEAT SHOCK PROTEINS MODULATE SIGNALING PATHWAYS IN SURVIVAL OF STRESSED FISH TO POLLUTED ENVIRONMENTS

Ekambaram Padmini *
Department of Biochemistry, Bharathi Women's College,
Chennai, Tamilnadu, India

ABSTRACT

Heat shock proteins are highly homologous chaperone proteins, present in all cells playing key roles in limiting the consequences of protein damage and facilitating its recovery. There are accumulating data about the involvement of HSPs in chaperoning function. However, the task of HSPs with regard to its participation in cell survival processes still remains undeciphered in response to stress stimuli in natural conditions. The objective of this review is to interpret the role of HSPs in stress tolerance and cell survival mechanisms in fish exposed to exposed to pollutant stress in natural field conditions. HSPs inhibit apoptosis by down regulating apoptotic events like apoptosome formation, release of apoptogenic factors and ASK1-mediated JNK1/2 signaling etc. On the other hand, HSP also favors survival by promoting the activities of pro survival kinases like ERK1/2 and Akt. Generally, the overall balance between the signaling pathways of survival and death determine the fate of the cell. HSPs modulate multiple events within the signaling pathways in stressed fish thereby helping them to sustain survival.

INTRODUCTION

Exposure to environmental stress may affect the cellular equilibrium. This condition disturbs normal cellular functions and has lethal consequences. Environmental stress includes

* Corresponding author details:Dr. E. Padmini,Associate Professor,Department of Biochemistry,Bharathi Women's College,Chennai-600 108, Tel: +91 44 26213748; Fax: +91 44 25280473,E-mail address: dstpadmini@rediffmail.com

hyperthermia, hypoxia, heavy metals, oxidative stress, tissue damage and infections. The impact of these stresses depends on its intensity and duration. The ability to respond appropriately to stress is essential for surviving in an ever changing environment. The activation of a cellular stress response characterized by the rapid synthesis of a group of highly conserved proteins called heat shock proteins helps the organisms to cope with stress.

Heat shock proteins belong to the superfamily of evolutionally conserved, intracellular proteins forming a part of the cellular defense. They are among the most highly and ubiquitously expressed proteins in all cells from bacteria to mammals (Csermely et al., 1998). They are responsible for diverse cellular processes like protein folding, activation, transport and oligomeric assembly (Csermely et al. 1998). HSPs principally function by preventing potentially damaging interactions. It aids in disassembling the formation of protein aggregates, thus playing a fundamental role in the maintenance of cellular homeostasis. They collectively serve as one of the molecular mechanisms that animals utilize to tolerate stress. The studies of the protective ability of the HSPs have focused largely on their role as chaperones to prevent misfolding and refolding or renaturation of proteins (Beere, 2004). However, the function of HSPs is broader and encompasses an anti-apoptotic role and these functions do not depend upon their chaperoning ability (Parcellier et al., 2003a). Hence these proteins can have pleiotrophic effects, by interacting with a very large variety of cellular proteins and thus are important components of cellular networks (Csermely, 2004).

Estuarine pollutants in recent years pose a major threat to aquatic organisms like fish. The ecotoxicants elicit their effects via redox cyclingproperty by the development of oxidative stress impacting fish health at various levels. In spite of this, fish that inhabit polluted estuaries are able to tolerate and resist the stress status, surviving successfully in that environment. Mammalian species are the usually used models to study oxidative stress and to clarify the mechanisms involved in cellular damage and response. Though few of the studies are carried out in fish, many of them are limited to *in vitro*conditions, neglecting to decipher the cellular mechanisms that actually happen in a natural environment against pollutant induced oxidative stress. As HSPs are regarded as the immediate responders to stress, the examination of the expression pattern of this protein and the signal transduction pathways associated with cellular stress response will allow identifying the regulatory mechanisms that enable fish to survive in a stressed environment. Hence the present review focuses on emerging information that describes the mechanisms and functional significance of HSPs in the modulation of signal transduction pathways facilitating towards the effect of cell survival.

Pollution Impact on Fish

Fish are exposed to multiple types of stressors in the natural environment. The well documented environmental stressor includes the environmental pollutants. For instance, pollutants like polycyclic aromatic hydrocarbons, polychlorinated biphenyls, arsenic, chlorine and cyanide serve as potent stressors for all salmonid species. Heavy metals such as chromium, manganese, iron, nickel, zinc, selenium, lead and mercury, the leading environmental metallic pollutants exert significant stress impact on estuarine grey mullet *M. cephalus* (Padmini and Usha Rani, 2009b; Padmini and Vijaya Geetha, 2007b). High concentrations of metals such as copper, cadmium, zinc, and iron may also cause death in exposed fish. Other potential environmental contaminants include insecticides, herbicides, fungicides and defoliants. Increasing human population and urbanization, as well as agricultural and industrial activities, contribute contaminants to the environment that can affect fish at all life stages (Dhaliwal and Kukal 2005). The major pollution impact on fish are observed via the redox cycling property of heavy metals and other contaminants which serves as the major contributors of oxidative stress. A state of unbalanced tissue oxidation characterized by elevated levels of free radicals and decreased levels of antioxidants were observed in fish exposed to pollutant stress (Padmini 2010). Those fish also demonstrated accumulation of damaged and oxidized macromolecules like lipid, proteins and DNA in various organs, decreased reproductive success, increased susceptibility to infection and shorter life spans associated with cell death (Padmini and Sudha 2004; Padmini et al., 2004).

HSPs-Saviour of Cells

A stressor can be any sudden change in the cellular environment to which the cell is not prepared to respond. Almost all types of cellular stressors induce HSPs and several different types of environmental stressors may trigger HSP overproduction. Highly expressed HSPs have been documented in fish cells in response to different types of stressors like hypoxia, heavy metals, PAH, detergents, food deprication, reduced oxygen levels, heat shock, hyperammonemic conditions ets. HSPs play a key role in various aspects of fish physiology, including development, aging, stress physiology, endocrinology, immunology, environmental physiology, adaptation and stress tolerance. It plays a role in the long-term adaptation of animals to their environment (Morimoto and Santaro 1998). A positive and direct correlation between HSP expression and thermotolerance has been well documented (Russotti et al., 1996). For example Deaney and Klesius (2004) reported the induction of HSP70 in the blood, brain and head-kidney of juvenile nile tilapia during exposure to hypoxic conditions. Cara et al. (2005) reported the impact of food deprivation, reduced oxygen levels and heat shock on the expression of HSP70 and HSP90 in the early life stages of the gilthead sea bream. Clarkson et al. (2005) have also reported that HSP90 induction is important after exhaustive exercise. HSP's induction plays important role in protection against cellular damage and in providing stress resistance.

Stress-Activated Signaling Pathways

Reactive oxygen species are derived from multiple sources, including environmental stress. Major ROS include superoxide, hydrogen peroxide and hydroxyl radical, as well as nitric oxide. ROS have been described as essential second messengers in several metabolic pathways and also in the various signal transduction cascades, leading to the activation of various kinases and transcription factors (Jacquier-Sarlin and Polla, 1996).

MAPK Pathways

Mitogen-activated protein pathways constitute a large kinase network that regulates a variety of physiological processes. MAPKs are involved in many aspects of development, disease and cellular responses to stress. MAPKs have been to shown to mediate a vast number of cellular responses including gene transcription, cytoskeletal organization, metabolic homeostasis, cell growth and apoptosis in response to many different intracellular signals (Kyriakis 1999). These kinases deserve potential importance in mediating organismal responses to multiple environmental stresses. In the large MAPK subfamily, three subgroups have been identified: the extracellular signal-regulated kinase 1/2 (ERK1/2), the c-Jun NH_2-terminal kinase 1/2 (JNK1/2) and p38 MAPK. They play major roles in the regulation of intracellular metabolism and gene expression(Cowan and Storey 2003). They are responsible for the phosphorylation of a variety of proteins including downstream kinases and transcription factors. ERK, JNK and p38 can all be activated by a variety of stimuli including growth factors, cytokines and different cellular stresses. However, the subfamilies are differentially affected by particular stimuli. In general, ERK shows greater activation than either JNK or p38 in response to mitogenic stimulation and has been associated with cell survival. In contrast, JNK and p38 are activated to a much greater extent by stressful stimuli and have been linked to cell death (Waskiewicz and Cooper 1995; Robinson and Cobb 1997).

ERK Pathway

The ERK pathway response primarily to growth factors and mitogens and stimulates tanscrptional responses in the nucleus. ERK1 and ERK2, are activated by MAPK/ERK kinase (MEK) 1 and MEK2 which phosphorylate at the Thr-Glu-Tyr motif (Cobb and Goldsmith, 1995). The MEKs, in turn, are activated by c-Raf, the MAPKKK of this signaling pathway that is turn regulated by growth factor receptors and tyrosine kinases activating through Ras (Moodie and Wolfman, 1994). Upon translocation to the nucleus, ERKs are responsible for the phosphorylation of multiple substrates, depending on the initial stimulus. These include activators of transcription including p90 RSK S6 kinase (Frodin and Gammeltoft, 1999), MAPK-activated protein kinase-1, MAPKAP-K1, phospholipase A2 and MSK, as well as transcription factors (Elk-1, Ets 1, Sap1a, m-Myc), STAT (signal transducers and activators of transcription) proteins such as Stat3, adapter proteins such as Sos, growth factor receptors such as epidermal growth factor (EGF), and the estrogen receptors (Denhardt, 1996). Generally, activation of an ERK signaling pathway has a role in mediating cell division, migration and survival. ERK1/2 and MEK1/2 are also strongly activated during muscle exercise and may provide the link between exercise and adaptive changes in skeletal muscle composition (Widegren et al., 2000). ERK is also shown to be highly activated by treatment with oxidants such as H_2O_2 and it has been shown to be important for long-tern survival after such treatment in NIH 3T3 mouse fibroblasts and rat PC12 phaeochromocytoma cells (Guyton et al., 1996).

JNK Pathway

JNK is activated in response to a variety of extracellular stimuli, including TNFα, H_2O_2, UV light, X-rays, heat shock and growth factor- or serum-withdrawal (Verheij *et al.* 1996; Lin 2003). Two MAP2Ks (JNKK1/MKK4/SEK1 and JNKK2/MKK7) for JNK have been identified (Liu *et al.* 2004). Several MAP3Ks, such as MAPK kinase kinase 1 (MEKK1), ASK1, MLK, TAK1 and TPL-2 have been reported to act as MAP3Ks for JNK and ASK1 is considered to be its most important activator (Lin 2003). Activation of JNK is also regulated by scaffold proteins such as JIP, β-arrestin, JSAP1, protein phosphatases and NF-κB (Liu *et al.* 2004).

The JNK family has two ubiquitously expressed isoforms, JNK1 and JNK2, and a tissue-specific isoform, JNK3, all of which have two different splicing forms, p54 and p46 (Davis 2000; Chang and Karin 2001; Lin 2003). JNK1 and JNK2 have been considered redundant isoforms, both of which contribute to cellular JNK activity in response to extracellular stimuli (Kallunki *et al.* 1994; Davis 2000). Gene disruption experiments suggest that despite some functional differences, JNK1 and JNK2 have overlapping functions (Sabapathy *et al.* 1999a, 1999b). Both JNK1 and JNK2 were shown as positive regulators of c-Jun by promoting its stability and thereby increasing AP1-dependent gene expression(Davis 2000). Biochemical experiments have shown that JNK2 has a 25-fold higher binding affinity for c-Jun than JNK1, leading to the suggestion that JNK2 is the major c-Jun kinase (Kallunki *et al.* 1994). However, other experiments have suggested that the JNK1 isoforms may be slightly more efficient in phosphorylating c-Jun (Gupta *et al.* 1996). Recently, microtubule-associated proteins were shown to be phosphorylated and regulated by JNK1 rather than JNK2 (Chang *et al.* 2003). It thus seems that the different JNK isozymes may have evolved for specific biological functions, probably depending on the activating stimuli and responding cell type.

JNK has been implicated in regulation of many cellular activities from gene expressionto programmed cell death (apoptosis) (Davis 2000; Chang and Karin 2001; Lin 2003). It is activated by stress signals, proinflammatory signals and some mitogenic signals (Davis 2000). It is also a key mediator of cell responses to environmental stimuli (Jaeschke *et al.* 2006). Upon stimulation, activated JNKs phosphorylate transcription factors like c-Jun, JunD and ATF2, which participate in the activation and formation of the AP-1 complex (Shaulian and Karin 2002). It also regulates the activities of Ets-like protein 1 (ELK1), p53 and c-Myc, as well as other factors, such as members of the Bcl-2 family (Yu *et al.* 2004). JNK activation was shown to either induce apoptosis or stimulate cellular proliferation and transformation (Weston and Davis 2002). Activation of JNK1/2 by ASK1 favors pathways that lead to apoptosis. Moreover, in the absence of NF-κB activation, prolonged JNK activation promotes TNFα induced apoptosis (Tang *et al.* 2001, 2002). JNK activation is also required for UV-induced apoptosis (Tournier *et al.* 2000). However, there is evidence that JNK also contributes to cell survival. JNK isoforms are considered as positive regulators of c-Jun expression and cellular proliferation (Jaeschke *et al.* 2006). Genetic evidence reveals that JNK1 and JNK2 are involved in survival of neuronal cells in mouse forebrain and hindbrain regions during development (Sabapathy *et al.* 1999b). Recently, it has been shown that JNK activation is required for IL-3-mediated survival via phosphorylation and inactivation of the proapoptotic Bcl-2 family protein BAD (Yu *et al.* 2004). Thus, the function of JNK in apoptosis depends on the cell type, nature of the apoptotic stimulus, duration of its activation and activity of other signaling pathways (Lin 2003).

p38 Pathway

p38 pathway is activated by environmental stresses, including heat, osmotic and oxidative stresses, ionizing radiation, ischemia as well as inflammatory cytokines and TNF receptor signaling (New and Han 1998). Hence several MAPKKKs can initiate the p38 MAPK signaling pathway and that the specificity of activation may be determined by the stimuli.The upstream kinases acting on p38 include MKKs that are in turn activated by MEKKs, MLKs and ASK1. The p38 MAPK protein is represented by four isoforms: α, β, γ and δ.Activation of these isoforms is achieved by dual phosphorylation of a threonine and a tyrosine within the threonine-glycine-tyrosine sequence in the activation domain of the kinase (Ashwell 2006). The p38 pathways plays an essential role in regulating the expression of many inflammatory molecules, differentiation of epidermal keratinocytes, myoblasts, and immune cells as well as mediates innate immune responses (Efimova et al., 2003; Ashwell 2006). It has been demonstrated in several animal models that activation of the p38 MAPK pathway is required for apoptosis induction (Li et al., 2003; Tanaka et al., 2002; Grethe and Porn-Ares 2006). However, the molecular mechanisms that determine whether p38 signaling either promotes or inhibits cell proliferation and survival have not been elucidated but could potentially be linked to the transformation state of the cell or could depend on the nature of p38-activating signal.

JAK-STAT Signaling Pathways

The Janus kinase/signal transducer and activator of transcription(JAK/STAT) pathway is one of the important pleiotropic cascadesused to transduce a multitude of signals for development andhomeostasis in animals. TheJAK/STAT pathway is the principal signaling mechanism that takes part in the regulation of cellular responses for awide array of cytokines and growth factor receptor tyrosine kinases and G-protein-linked heptahelical receptors (Ihle *et al.* 1994). The pathway transduces the signal carried by these extracellularpolypeptides to the cell nucleus, where activated STAT proteins modify gene ex-pression. Although STATs were originally discovered as targets of JAKs, it has now become apparent that certain stimuli can activate them independently of JAKs. JAK/STAT activation stimulatescell proliferation, differentiation, cell migration and apoptosis (Rawlings *et al.* 2004).

Phosphoinositide 3-kinase (PI3K)/Akt Signaling Pathway

The serine/threonine kinase Akt, also known as protein kinase B (PKB) is originally identified as the cellular homolog of the v-akt oncogene. It is activated via a PI3K dependent signaling pathway when cells or tissues are exposed to growth factors, insulin, certain cytokines and other extracellular stimuli. PI3K reside mainly in the cytosol until recruited into active signaling complexes at the plasma membrane, where they are involved in the generation of 3'-phosphorylated phosphoinositides that function as signaling intermediates in signal transduction cascades. Akt, the target of PI3K is associated with the inhibition of apoptosis in a variety of ways (Datta *et al.* 1997). It has received widespread attention as an important anti-apoptotic protein through which various survival signals suppress cell death induced by growth factor withdrawal, detachment of cells from their extracellular matrix and other stress stimuli (Wang *et al.* 2000). The PI3K/Akt pathway transduces survival signals through phosphorylation processes and regulates pro- and anti-apoptotic factors such as B-cell lymphoma 2 (Bcl-2)-associated death promoter (BAD), caspase-9 and IκB kinase (IKK). It has been reported that Akt protects against mitochondrial disruption and apoptosis by activating ERK1/2 and NF-κB and inhibiting JNK and Bax phosphorylation (Kim *et al.* 2001).

Nuclear Factor Kappa B (NF-κB) Signaling Pathway

NF-κB is an important regulator of the balance between cell survival and cell death. Hence it is presumed to have an important role in determining cell fate during stress. NF-κB is activated by inflammatory cytokines [tumor necrosis factor (TNF)-α and interleukin (IL)-1β], oxidative stress, protein kinase C (PKC) and PI3K (Conde de la Rosa *et al.* 2008). A heterodimeric transcription factor, NF-κB leads to the transcriptional activation of numerous stress response genes (Baeuerle and Henkel 1994). Interestingly, both the ERK and the Akt signaling pathways have been implicated in NF-κB activation through phosphorylation of inhibitory κB (IκB). MAPK and Akt signaling pathways and NF-κB activation have been demonstrated in many studies in regulating the cellular response to stress (Hirano *et al.* 1996; Malinin *et al.* 1997; Schouten*et al.* 1997). NF-κB signaling pathway has been described to antagonize hepatocyte cell death by influencing the balance between pro- and anti-apoptotic signals. It has been postulated that NF-κB inhibits TNFα-induced accumulation of ROS that normally mediate prolonged JNK activation and cell death (Sakon *et al.* 2003). Indeed, inhibition of NF-κB activity induces apoptosis in hepatocytes, suggesting its role in the transcription of anti-apoptotic genes (Bellas *et al.* 1997).

HSPs in Signal Transduction

HSPs are the evolutionarily conserved protein families in the eukaryotic kingdom. The HSP70 family represents one of the most widely examined heat shock protein families and the members of the HSP70 family differ in their spatial and subcellular distribution, as well as in their expression levels under normal and stressed conditions (Rohde *et al.,* 2005). HSPs like HSP90 serves as the key components in the cellular multi-chaperone complexes in non-stressed cells where it performs house keeping functions to control the activity, turnover and trafficking of a variety of "client" proteins (Pratt and Toft, 2003). It has been suggested that HSP90 specifically chaperones the protein kinases in the signal transduction pathways and the transcription factors and has been known as "specific molecular chaperone". The molecular relationships between heat shock proteins and various signaling proteins appear to be critical for the normal function of cellular signal transduction pathways (Basso et al., 2002; Fujita et al., 2002; Sato et al., 2000).

The signaling pathways are all cascades that are susceptible to regulatory inputs at multiple levels within the cascade as well as *via* multiple mechanisms. These can include cell/tissue specific expression patterns, specificity of the stimuli that can trigger each pathways, specificity of the substrates that acts as targets, modification of responses *via* activators, inhibitors, scaffolding proteins, sequential interactions with other proteins in a cascade, positive and negative feedback loops, and cross-talk among different signaling pathways.

Stress response, the cellular protective and adaptive response associated with the induction of antioxidant proteins like HSPs are the first discovered in the face of an oxidizing environment. They represent one of the highly important survival pathways that help to cope with stress-induced conformational damage of proteins maintaining cellular homeostasis (Papp *et al.* 2003).These HSPs are induced during a variety of stressful conditions including heat shock, hyperosmotic stress, oxidative stress, UV radiation, amino acid analogues, infection, inhibitors of energy metabolism and heavy metals (Morimoto *et al.,* 1992).

HSP70 is essential for protein folding, translocation across cellular compartments, assembling and maintaining multi-protein complexes in active states, and preventing self-association. It also directs misfolded and short-lived proteins to destruction by the proteasome in order to maintain protein homeostasis (Kelley and Georgopoulos 1992). HSP70 is cell cycle regulated and is found to be upregulated to assist in the protein refolding under stressed condition (Guo et al. 2007) by their enhanced peptide-binding ability and peptide complex stability (Callahan et al. 2002).Mitochondrial heat shock protein 70 (mtHSP70) is found to play a primary role in cellular defense against physiological stress like exposure to environmental contaminants and helpful in the maintenance of cellular homeostasis by promoting the cell survival. The authors have proved that increased level of HSP70 appear to be a major contributor to the acquired tolerance against pollution stress and the adaptation of fish in contaminated estuary (Padmini and vijayageetha ,2009).

HSP90, the most abundant chaperone in eukaryotic cells is involved in multiple and biological processes such as cell proliferation, differentiation and apoptosis (Terry et al., 2005, Lanneau et al., 2008). The inducible isoform HSP90α is essential for maturation and activation of proteins. It plays appreciable role in stress tolerance, protein folding and regulation of signal transduction pathways that control cell growth and survival (Parsell *et al.* 1993). Its role in the regulation of cell survival and death pathways via the modulation of the activity of a vast number of its client proteins is highly documented (Picard 2002; Sreedhar *et al.* 2004).

The overexpression of HSP90α under oxidative stress condition has been demonstrated earlier in test fish hepatocytes; the upregulation being mediated via the transcriptional induction of HSP90α genes by heat shock factor 1 (HSF1) (Padmini and Usha Rani 2009a). These authors have also demonstrated that significant induction of HSP90α serves as an adaptive strategy in response to pollutant related oxidative stress in hepatocytes of fish from test site compared to control counterpart (Padmini and Usha Rani, 2010).

HSPs Inhibit Apoptotic Signaling Pathways

HSPs have a role to play in long-term adaptation and increased stress tolerance (Morimoto and Santoro 1998). HSP70 performs a pivotal role in the process of apoptotic tolerance by binding to specific target proteins mediating their cellular activities. HSP inhibits cell apoptosis by inhibiting the proapoptotic kinase, ASK1 (Zhang *et al.* 2005). HSP70 inhibits apoptosis through either its chaperoning function or its binding activities to specific target molecule. A direct association of HSP 70 with ASK1 is reported to be critical for the regulation of ASK1 activity (Sato *et al.* 2000; Basso *et al.* 2002, Park et al. 2002). HSP70 can control Ask mediated apoptosis indirectly by upregulation of Trx, which binds to ASK-1 in a reduced form. The Trx-ASK1 complex subjects the bound ASK1 to ubiquitination and proteasomic attack (Berggren et al., 1996). Pollution stress induced HSP70 overexpression noted in stressed fish liver mitochondria mitigates stress induced ASK1 expression by elevating Trx expression (Padmini et al., 2009). The role of HSP70 in upregulating the expression of Bcl-2 protein in stressed fish liver mitochondria has also been demonstrated to block mitochondria mediated apoptosis. The role of JNK in activating Bcl-2 has also been analyzed tocontrol apoptosis (Srivastava et al., 1999).

The inhibitory effect of Hsp90 on ASK1–p38 activities is diminished when the Akt phosphorylation site on ASK1 (pSer83) is absent or when Akt is genetically deleted in cells, suggesting that Hsp90 and Akt function together to inhibit ASK1–p38 signaling (Zhang *et al.* 2005). Recent study has also suggested that HSP90α inhibits oxidative stress induced ASK 1 activation in hepatocytes of fish from a polluted estuary compared to unpolluted counterpart (Padmini and Usha Rani, 2010). Over expression of HSP90α contributes to retain ASK1 in an inactive state, thereby inhibiting stress-induced ASK1 signaling and hepatocyte apoptosis.

JNK is the downstream target of ASK1, and its activation in response to stimuli such as ROS, proinflammatory and mitogenic signals is generally involved in stress response (Deng *et al.* 2003). It has been suggested that MAPKs such as JNK are redox sensitive and may be the effector molecule in stress signals induced apoptosis (Kyriakis and Avruch 2001; Shen and Liu 2006).It has been well established that oxidative stress can upregulate JNK by activating its upstream kinase such as ASK1 (Shen and Liu 2006). However, it has also been suggested that JNK activity is regulated during inhibition of ASK1 by its interacting partners (Saitoh *et al.* 1998; Kim *et al.* 2001). Upregulated HSP 70 and 90α under stress condition, associate with ASK1 and terminate ASK1 mediated JNK1/2 signaling cascades.

Activation of Prosurvival Pathways by HSPs

The role of HSP70 in cell survival has been associated with its chaperoning and non-chaperoning function. In stressed condition HSP70 can chaperone various cell survival proteins in-order to restore protein homeostasis and will aid in cell survival. Under extreme stress HSP70 interacts with various signaling molecules to reduce apoptosis and maintain cell viability. Sreedhar and Csermely (2004) reported that the JNK is stress responsive as HSP70 interacts directly with peptide-binding domains of JNK. JNK signal transduction pathway has been found to be involved in heat shock gene expression (Lee and Corry, 1998). AsHSP70 is thought to protect cells against cellular stress, the adaptive role of elevated levels of HSP70 in

Ennore estuary fish liver mitochondria was correlated with JNK1 and JNK2 expression levels, suggests that mtHSP70 showed a significant positive correlation with JNK1 as it is meant for cell proliferation. Our study results show that JNK1 activation and upregulation of HSP70 serves as an intervening checkpoint prior to the commitment of apoptosis in stressed fish liver mitochondria (unpublished data).

Akt, a major survival kinase, was found to be under the regulation of Hsp70, and when the ATPase activity of Hsp70 was increased or decreased, Akt levels were also increased or decreased suggesting that HSP70 has the ability to interact with Akt and results in their stabilization (Koren et al., 2010). ThusHSP70 can promote Akt mediated survival mechanism (Gao et al., 2002). The interaction of HSP70 and ERK has not been well demonstrated, however the role of BAG-1 (Bcl-2 associated antogene), a molecular chaperone regulator in the activation of Raf proteins and thereby resulting in the activation of ERK has been demonstrated (Wang et al., 1996) Raf-1 and HSP70 may compete for binding to BAG1 (Song et al., 2001). When levels of HSP70 are elevated after cell stress, the BAG1_Raf-1 complex is replaced by BAG1_Hsp70, (Briknarova et al., 2002) there by contributing to cell survival..

HSP90α is critical for maintaining normal cellular homeostasis (Maloney and Workman 2002). It plays an important role in facilitating the proper folding, maturation and activity of its client proteins (Neckers 2002).JAK1/2, the upstream activator of STAT, is one of the important client proteins of HSP90. Hence JAK1/2 has been described to interact with HSP90 and could be involved in a permanent JAK-STAT signaling (Shang and Tomasi 2006). Recently HSP90 has also been reported to interact with STATs and this interaction is required for STAT phosphorylation, further hypothesizing that STAT 1α and 3 might also act as a client of HSP90 (Shang and Tomasi 2006). In accordance with these reports, a recent study has suggested significant activation of STAT molecules during HSP90α upregulation indicating a crucial for HSP90α plays in STAT signaling (Padmini and Usha Rani, Unpublished data). It has been demonstrated that inhibition of HSP90 could be associated with the interruption of permanent STAT activation (Schoof *et al.* 2009). Treatment of classical Hodgkin lymphoma (cHL) cell line with 17-AAG, led to reduced cell proliferation and a complete inhibition of STAT phosphorylations probably as a result of reduced protein expression of JAKs. A reduction of STAT phosphorylations along with decreased cell proliferation of L428 cells was also visible after knock-down of HSP90 (Schoof *et al.* 2009), the study suggesting that HSP90 is important for the permanent activation of STAT molecules.

ERK1/2 and Akt of the Ras/Raf/MAP kinase cascade and PI3K pathway respectively are the two key prosurvival kinases being implicated in influencing cell survival in response to stress (Helmreich 2001). They are best known for their high activation in response to growth factor stimulation. They play important roles in regulating cell proliferation and preventing apoptosis (Kandel and Hay 1999). A direct correlation between oxidative stress resistance and ERK1/2 or Akt activation has also been put forth by Ikeyama *et al.* (2002). It has also been shown that ERK and PI3-K/Akt signaling pathways contribute to survival of cells following their treatment with H_2O_2 (Guyton *et al.* 1996; Wang *et al.* 2000). HSP90α has been reported to regulate Raf, the upstream kinase of ERK1/2 in the Ras-Raf-MEK-ERK pathway and Akt of PI3K pathway (Schulte *et al.* 1995; Meares *et al.* 2004). Hence HSP90α plays role in the survival pathways mediated by ERK1/2 and Akt. Disruption of HSP90α function inhibits the interaction of this chaperone with its client proteins (Beck *et al.* 2009; Shen *et al.* 2009). Also, loss of protection provided by HSP90 facilitates their degradation (Dou *et al.* 2005). It is demonstrated to act as a molecular chaperone in late-phase activation of ERK1/2 stimulated by oxidative stress in vascular smooth muscle cells (Liu *et al.* 2007). Piatelli *et al.* (2002) have also reported inhibition of ERK1/2 activation during disruption of HSP90 function. Similarly, Akt is complexed with HSP90 and this interaction is necessary for this protein to perform various processes in cellular signaling (Sato *et al.* 2000; Fontana *et al.* 2002; Fujita *et al.* 2002). Consequently, many reports have verified that Akt levels are downregulated and Akt dependent survival activities are suppressed following inhibition of HSP90 with geldanamycin (Basso *et al.* 2002; Doong *et al.* 2003; Meares *et al.* 2004). A significant activation of ERK1/2 and Akt has also been observed during HSP90α upregulation in fish hepatocytes inhabiting polluted estuary demonstrating the positive regulatory role of HSP90α on these prosurvival kinases. Dou *et al.* (2005) have also reported that the phosphorylation levels of ERK in primary neurons decreased after treatment with geldanamycin, while the total protein level of ERK did not change, indirectly suggesting the regulation of ERK activity by HSP90. In the same way, it is reported using transfection experiments that Akt activation parallels HSP90 induction and the formation of the Akt-HSP90 complex stabilizes Akt protecting the cells from undergoing apoptosis (Sato *et al.* 2000).

Regulations of cell growth are dependent on a number of gene families including proto-oncogene, growth factor, growth factor receptor and immediate early transcription factor gene. ELK1 is known to be involved in the regulation of immediate-early genes such as *c-fos* upon mitogen activation, and thus commonly implied in cell proliferation (Demir and Kurnaz 2008). It plays an important role in transducing extracellular signals into a nuclear response by acting as targets for the MAPK signaling pathways like ERK1/2 and JNK1/2 (Sharrocks 2002). It has been reported that ERK1/2 activation by mitogenic stimulation phosphorylates ELK1, increasing its affinity for the SRF and enhancing transcription of growth related proteins (Aplin *et al.* 2001).

Bcl-2, an apoptosis suppressing protein can inhibit apoptosis induced by a variety of stimuli including oxidative stress (Kaufmann *et al.* 2003). This protein under non-stressed condition is in complex with a proapoptotic protein, BAD. Upon a growth or survival stimuli, the interaction between these proteins is regulated primarily by Akt. Akt phosphorylates BAD on a serine, which causes its release from the complex with Bcl-2, allowing the latter to perform its anti-apoptotic function (Datta *et al.* 1997; Franke and Cantley 1997). Several recent studies have provided evidence suggesting that Bcl-2 exerts its anti-apoptotic function through suppression of the JNK signaling pathway. For example, overexpression of Bcl-2 was found to promote survival and to block JNK activation caused by withdrawal of nerve growth factor in PC12 cells (Park *et al.* 1996). Bcl-2 upregulation was also demonstrated to prevent JNK activation and to suppress apoptosis of N18TG neuroglioma cells caused by a variety of agents (Park *et al.* 1997). Although JNK overexpression was able to antagonize this effect of Bcl-2, it did not seem to act directly on JNK to inhibit its activity because the activation of MEKK1, an upstream intermediate of the JNK cascade, was also prevented by Bcl-2 overexpression (Park *et al.* 1997). It has been shown that stimulation of cardiac myocytes with 0.2 mM H_2O_2, which induces apoptosis, resulted in a marked down-regulation of Bcl-2 protein (Markou *et al.* 2009). Dunschede *et al.* (2008) have reported that Bcl-2 upregulation might protect against the postischemic burst of ROS and therefore reduces apoptotic-related cell death. Oxidative stress-dependent upregulation of Bcl-2 expression has also been demonstrated in the central nervous system of aged rats (Kaufmann *et al.* 2003). A cooperative control between Akt activation and Bcl-2 expression has been reported in capillary endothelial cells (Flusberg *et al.* 2001). Pugazhenthi *et al.* (2000) have also defined upregulation of Bcl-2 expression as a novel anti-apoptotic function of Akt signaling.

Among the potential MAPK-regulated transcription factors known to be activated in response to oxidative stress is NF-κB. Under normal growth conditions, NF-κB is present in the cytoplasm in an inactive state and translocates from cytosol to nucleus upon activation by a wide variety of stimuli including the proinflammatory cytokine TNFα, oxidative stress, bacterial and viral proteins (Ahn and Aggarwal 2005). Bellas *et al.* (1997) have suggested that active NF-κB has important role in the control of cell proliferation and survival. Inhibition of this protein activation results in increased susceptibility to apoptosis leading to cell death. During environmental contaminant mediated oxidative stress, HSP90α mediated enhanced activation of NF-κB characterized by its translocation from cytosol to nucleus was observed suggesting the prevalence of prosurvival mechanism mediated by NF-κB in Ennore test fish hepatocytes (Padmini and Unpublished data). A recent finding has also reported a significant increase in NF-κB activation in hyperoxia model of oxidative stress (Li *et al.* 1997).

NF-κB in the cytosol is kept non-active through its interaction with one or more members of the IκB family of inhibitory proteins. Conversion of NF-κB to an active form requires the release of IκB. This is accomplished through the phosphorylation of IκB leading to its ubiquitination and degradation. The ERK signaling pathway has been linked to NF-κB activation through the finding that the ERK regulated kinase p90-RSKcan phosphorylate IκB, leading to its inactivation in response to mitogenic stimulation (Schouten *et al.* 1997). The same pathway also downregulates the expression of PAR-4, an inhibitor of NF-κB activation (Barradas *et al.* 1999). Akt activates NF-κB via regulating IKK, thus result in NF-κB mediated transcription of pro-survival genes(Faissner *et al.* 2006). Hence HSP90α mediated ERK1/2 and Akt activations signals for survival also via NF-κB activation.

Junttila *et al.* (2008) have also depicted the function of the ERK pathway as survival-promoting, in essence by opposing the proapoptotic activity of the stress-activated JNK pathways. During apoptosis, the ERK mediated survival signaling is inhibited through JNK induced PP2A-mediated inhibition of MEK1, 2 (the upstream activator of ERK1/2). Many studies have also demonstrated that ERK and JNK pathway activities oppose each other as a means of regulating apoptosis and under conditions of severe stress, activated JNK suppress the survival-promoting activity of the ERK pathway (Black *et al.* 2002; Shen *et al.* 2003; Friedman and Perrimon 2006). Similarly, active Akt in association with HSP90 is reported to inhibit proapoptotic kinase ASK1 activity. HSP90-Akt binds to and phosphorylates ASK1 at serine 83 to maintain ASK1 in an inactive state. Inhibition of HSP90 or PI3K-Akt signaling disrupts the HSP90-Akt-ASK1 complex leading to activation of ASK1 signaling (Zhang *et al.* 2005).

CONCLUSION

In normal physiological contexts, all the signaling pathways function independently with no crosstalk between them, whereas interplay between pathways would be induced during stressful situations when signal strength exceeds the capacity of the pathway. The crosstalk observed between cellular mechanisms and signaling pathways is to increase the specific cellular responses against pollutant stress. Induction of HSP70 and HSP90α followed by activation of cell signaling pathways (STAT, ERK1/2, Akt, JNK and NF-κB) regulated by these HSPs mediate hepatocellular resistance to oxidative stress induced by environmental pollutants facilitating their survival. At the same time, HSP70-Bcl-2-Trx pathway and HSP90-Akt pathway acts to down regulate ASK1 allowing inhibition of ASK1-mediated apoptotic signaling. Thus a crosstalk between anti and pro-apoptotic signaling pathways regulated by HSP70 and HSP90α modulate the balance between survival and death, ultimately favoring the shift towards oxidative stress tolerance and cellular survival. Hence it is conceivable thatinduction of HSPs in response to pollutant stress provides an adaptational strategy by favouring signal transduction mechanisms that work towards survival (STAT, ERK1/2-ELK1, Akt-Bcl-2 and NF-κB signaling) and against apoptosis (ASK1-JNK1/2 signaling).

References

Ahn KS, Aggarwal BB, 2005. Transcription factor NF-κB: a sensor for smoke and stress signals. *Ann NY Acad Sci* 1056: 218-233.

Ashwell JD, 2006. The many paths to p38 mitogen-activated protein kinase activation in the immune system. Nat. Rev. 6, 532-540.

Baeuerle PA, Henkel T, 1994. Function and activation of NF-κB in the immune system. *Annu Rev Immunol*12: 141-179.

Barradas M, Monjas A, Diaz-Meco MT, Serrano M, Moscat J, 1999. The downregulation of the pro-apoptotic protein Par-4 is critical for Ras-induced survival and tumor progression. *EMBO J* 18: 6362-6369.

Basso AD, Solit DB, Chiosis G, Giri B, Tsichlis P, Rosen N, 2002. Akt forms an intracellular complex with heat shock protein 90 (HSP90) and Cdc37 and

Beere HM, Wolf BB, Cain K, Mosser DD, Mahboubi A, Kuwana T, Tailor P, Morimoto RI, Cohen GM, Green DR, 2000. Heat shock protein 70 inhibits apoptosis by preventing recruitment of procaspase-9 to theApaf-1 apoptosome. *Nat Cell Biol* 2: 469-475.

Bellas RE, FitzGerald MJ, Fausto N, Sonenshein GE, 1997. Inhibition of NF-κB activity induces apoptosis in murine hepatocytes. *Am J Pathol* 151: 891-896.

Brazil DP, Hemmings BA, 2001. Ten years of protein kinase B signalling: a hard Akt to follow. *Trends Biochem Sci* 26: 657-664.

Chang L, Jones Y, Ellisman MH, Goldstein LS, Karin M, 2003. JNK1 is required for maintenance of neuronal microtubules and controls phosphorylation of microtubule-associated proteins. *Dev Cell*4: 521-533.

Chang L, Karin M, 2001. Mammalian MAP kinase signaling cascades. *Nature* 410: 37-40.

Cobb M, Goldsmith EJ, 1995. How MAP kinases are regulated. *J Biol Chem* 270: 14843-14846.

Cowan KJ, Storey KB, 2003. Mitogen-activated protein kinases: new signaling pathways functioning in cellular responses to environmental stress.*J Exp Biol* 206: 1107-1115.

Cowan KJ, Storey KB, 2003. Mitogen-activated protein kinases: new signaling pathways functioning in cellular responses to environmental stress.*J Exp Biol* 206: 1107-1115.

Csermely P, 2004. Strong links are important, but weak links stabilize them. *Trends Biochem Sci* 29: 331-334.

Csermely P, Schnaider T, Soti Cs, Prohaszka Z, Nadai G, 1998. The 90-kDa molecular chaperone family: structure, function and clinical applications. A comprehensive review. *Pharmacol Ther* 79: 129-168.

Datta SR, Dudek H, Tao X, Masters S, Fu H, Gotoh Y, Greenberg ME, 1997. Akt phosphorylation of BAD couples survival signals to the cell-intrinsic death machinery. *Cell* 91: 231-241.

Davis RJ, 2000. Signal transduction by the JNK group of MAP kinases. *Cell* 103: 239-252.

Deng Y, Ren X, Yang L, Lin Y, Wu X, 2003. A JNK-dependent pathway is required for TNFα-induced apoptosis. *Cell* 115: 61-70.

Denhardt, D. T. (1996). Signal-transducing protein phosphorylation cascades mediated by Ras/Rho proteins in the mammalian cell: the potential for multiplex signalling. *Biochem. J.* 318, 729-747.

Doong H, Rizzo K, Fang S, Kulpa V, Weissman AM, Kohn EC, 2003.CAIR-1/BAG-3 abrogates heat shock protein-70 chaperone complex-mediated protein degradation: accumulation of polyubiquitinated Hsp90 client proteins. *J Biol Chem* 278: 28490-28500.

Dou F, Yuan L-D, Zhu J-J, 2005. Heat shock protein 90 indirectly regulates ERK activity by affecting Raf protein metabolism. *Acta Biochimicaet Biophysica Sinica* 37: 501-505.

Dunschede F, Tybl E, Kiemer AK,Dutkowski P, Erbes K, Kircher A, Gockel I, Zechner U, Schad A, Lang H, Junginger T, Kempski O, 2008. Bcl-2 upregulation after 3-nitropropionic acid preconditioning in warm rat liver ischemia. *Shock* 30: 699-704.

Efimova, T., Broome, A. M., Eckert, R. L. (2003) A regulatory role for p38 delta MAPK in keratinocyte differentiation. Evidence for p38 delta-ERK1/2 complex formation. *J. Biol. Chem.* 278, 34277-34285

Efimova, T., Broome, A. M., Eckert, R. L. (2004) Protein kinase Cdelta regulates keratinocyte death and survival by regulating activity and subcellular localization of a p38delta-extracellular signal-regulated kinase 1/2 complex. *Mol. Cell. Biol.* 24,8167-8183

Faissner A, Heck N, Dobbertin A, Garwood J, 2006. DSD-1-Proteoglycan /Phosphacan and receptor protein tyrosine phosphatase-beta isoforms during development and regeneration of neural tissues. *Adv Exp Med Biol*557: 25-53.

Flusberg DA, Numaguchi Y, Ingber DE, 2001. Cooperative control of Akt phosphorylation, Bcl-2 expression, and apoptosis by cytoskeletal microfilaments and microtubules in capillary endothelial cells.*Mol Biol Cell* 12: 3087-3094.

Fontana J, Fulton D, Chen Y, Fairchild TA, McCabe TJ, Fujita N, Tsuruo T, Sessa WC, 2002. Domain mapping studies reveal that the M domain of hsp90 serves as a molecular scaffold to regulate Akt-dependent phosphorylation of endothelial nitric oxide synthase and NO release.*Circ Res* 90: 866-873.

Franke TF, Cantley LC, 1997. A Bad kinase makes good. *Nature* 390: 116-117.

Friedman A, Perrimon N, 2006. A functional RNAi screen for regulators of receptor tyrosine kinase and ERK signaling. *Nature* 444: 230-234.

Fujita N, Sato S, Ishida A, Tsuruo T, 2002. Involvement of HSP90 in signaling and stability of 3-phosphoinositide-dependent kinase-1. *J Biol Chem*277: 10346-10353.

Gabai VL, Sherman MY, 2002. Interplay between molecular chaperones and signaling pathways in survival of heat shock. *J Appl Physiol* 92: 1743-1748.

Gao T, Newton AC, 2002. The turn motif is a phosphorylation switch that regulates the binding of HSP70 to protein kinase C. *J Biol Chem.*277:31585–92

Grethe, S., Porn-Ares, M. I. (2006) p38 MAPK regulates phosphorylation of Bad via PP2A-dependent suppression of the MEK1/2-ERK1/2 survival pathway in TNF-alpha induced endothelial apoptosis. *Cell. Signal.*18,531-540.

Gupta S, Barrett T, Whitmarsh AJ, Cavanagh J, Sluss HK, Derijard B, Davis RJ, 1996. Selective interaction of JNK protein kinase isoforms with transcription factors. *EMBO J* 15: 2760-2770.

Guyton KZ, Liu Y, Gorospe M, Xu Q, Holbrook NJ, 1996. Activation of mitogen-activated protein kinase by H_2O_2. Role in cell survival following oxidant injury. *J Biol Chem* 271: 4138-4142.

Helmreich EJM, 2001. Components of signaling networks: linkers and regulators. In: The Biochemistry of Cell signaling. Oxford University Press, New York.

Hirano M, Osada S-i, Aoki T, Hirai S-i, HosakaM, Inoue J-i, Ohno S, 1996. MEK kinase is involved in tumor necrosis factor α-induced NF-κB activation and degradation of IκB-α. *J Biol Chem*271: 13234-13238.

Hirota K, Matsui M, Iwata S, Nishiyama A, Mori K, Yodoi J, 1997. AP-1 transcriptional activity is regulated by a direct association between thioredoxin and Ref-1. *Proc Natl Acad Sci USA* 94: 3633-3638.

Ihle JN, Witthuhn BA, Quelle FW, Yamamoto K, Thierfelder WE, Kreider B, Silvennoinen O, 1994. Signaling by the cytokine receptor superfamily: JAKs and STATs. *Trends Biochem Sci* 19: 222-227.

Ikeyama S, Kokkonen G, Shack S, Wang X-T, Holbrook NJ, 2002. Loss in oxidative stress tolerance with aging linked to reduced extracellular signal-regulated kinase and Akt kinase activities. *FASEB J* 16: 114-116.

Jacquier-Sarlin MR, Polla BS, 1996. Dual regulation of heat-shock transcription factor (HSF) activation and DNA-binding activity by H_2O_2: role of thioredoxin. Biochem J. 318,187-193.

Jacquier-Sarlin MR, Polla BS, 1996. Dual regulation of heat-shock transcription factor (HSF) activation and DNA-binding activity by H_2O_2: role of thioredoxin. *Biochem J* 318: 187-193.

Jaeschke A, Karasarides M, Ventura J-J, Ehrhardt A, Zhang C, Flavell RA, Shokat KM, Davis RJ, 2006. JNK2 is a positive regulator of the cJun transcription factor. *Mol Cell* 23: 899-911.

Jaeschke A, Karasarides M, Ventura J-J, Ehrhardt A, Zhang C, Flavell RA, Shokat KM, Davis RJ, 2006. JNK2 is a positive regulator of the cJun transcription factor. *Mol Cell* 23: 899-911.

Junttila MR, Li S-P, Westermarck J, 2008. Phosphatase-mediated crosstalk between MAPK signaling pathways in the regulation of cell survival. *FASEB J* 22: 954-965.

Kallunki T, Su B, Tsigelny I, Sluss HK, Derijard B, Moore G, Davis R, Karin M, 1994. JNK2 contains a specificity-determining region responsible for efficient c-Jun binding and phosphorylation. *Genes Dev* 8: 2996-3007.

Kallunki T, Su B, Tsigelny I, Sluss HK, Derijard B, Moore G, Davis R, Karin M, 1994. JNK2 contains a specificity-determining region responsible for efficient c-Jun binding and phosphorylation. *Genes Dev* 8: 2996-3007.

Kandel ES, Hay N, 1999. The regulation and activities of the multifunctional serine/threonine kinase Akt/PKB. *Exp Cell Res* 253: 210-229.

Kaufmann T, Schlipf S, Sanz J, Neubert K, Stein R, Borner C, 2003. Characterization of the signal that directs Bcl-X_L, but not Bcl-2, to the mitochondrial outer membrane. *J Cell Biol* 160: 53-64.

Kelly, W. L. and Georgopoulos, C. (1992). Chaperones and protein folding. *Curr. Opin. Cell Biol.* 4, 984-991.

Kim AH, Khursigara G, Sun X, Franke TF, Chao MV, 2001. Akt phosphorylates and negatively regulates apoptosis signal-regulating kinase 1. *Mol Cell Biol* 21: 893-901.

Koga F, Xu W, Karpova TS, McNally JG, Baron R, Neckers L, 2006. Hsp90 inhibition transiently activates Src kinase and promotes Src-dependent Akt and Erk activation. *Proc Natl Acad Sci USA*103: 11318-11322.

Koren J, Jinwal UK, Jin Y, O'Leary J, Jones JR, Johnson AG, Blair LJ, Abisambra JF, Chang L, Miyata Y, Cheng AM, Guo J, Cheng JQ, Gestwicki JE, Dickey CA, 2010. Facilitating Akt Clearance via Manipulation of Hsp70 Activity and Levels. *J Biol Chem*285 (4): 2498–2505,

Kyriakis JM, Avruch J, 2001. Mammalian mitogen-activated protein kinase signal transduction pathways activated by stressandinflammation. Physiol Rev 81: 807-869.

Lanneau D, Brunet M, Frisan E, Solary E, Fontenay M, Garrido C. Heat shock proteins: essential proteins for apoptosis regulation. *J Cell Mol Med.* 2008;12:743–761

Lee FS, Hagler J, Chen ZJ, Maniatis T, 1997. Activation of the IκBα kinase complex by MEKK1, a kinase of the JNK pathway. *Cell*88: 213-222.

Li Y, Zhang W, Mantell LL, Kazzaz JA, Fein AM, Horowitz S, 1997. Nuclear factor-κB is activated by hyperoxia but does not protect from cell death.*J Biol Chem*272: 20646-20649.

Li, S.-P., Junttila, M. R., Han, J., Kähäri, V.-M., Westermarck, J. (2003) p38 Mitogen-activated protein kinase pathway suppresses cell survival by inducing dephosphorylation of mitogen-activated protein/extracellular signal-regulated kinase kinase1,2. *Cancer Res.*63,3473-3477.

Lin A, 2003. Activation of the JNK signaling pathway: breaking the brake on apoptosis. BioEssays 25: 1-8.

Liu Dh, Yuan Hy, Cao Cy, Gao Zp, Zhu By, Huang Hl, Liao Df, 2007. Heat shock protein 90 acts as a molecular chaperone in late-phase activation of extracellular signal-regulated kinase 1/2 stimulated by oxidative stress in vascular smooth muscle cells. *Acta Pharmacologica Sinica* 28: 1907-1913.

Liu H, Nishitoh H, Ichijo H, Kyriakis JM, 2000. Activation of apoptosis signal regulating kinase 1 (ASK1) by tumor necrosis factor receptor-associated factor 2 requires prior dissociation of the ASK1 inhibitor thioredoxin.*Mol Cell Biol* 20: 2198-2208.

Liu J, Minemoto Y, Lin A, 2004. c-Jun N-terminal protein kinase 1 (JNK1), but not JNK2, is essential for tumor necrosis factor alpha-induced c-Jun kinase activation and apoptosis. *Mol Cell Biol* 24: 10844-10856.

Liu Y, Min W, 2002. Thioredoxin promotes ASK1 ubiquitination and degradation to inhibit ASK1-mediated apoptosis in a redox activity-independent manner. *Circ Res* 90: 1259-1266.

Malinin NL, Boldin MP, Kovalenko AV, Wallach D, 1997. MAP3K-related kinase involved in NF-κB induction by TNF, CD95 and IL-1. *Nature*385: 540-544.

Maloney A, Workman P, 2002. Hsp90 as a new therapeutic target for cancer therapy: the story unfolds. *Expert Opin Biol Ther* 2: 3-24.

Markou T, Dowling AA, Kelly T, Lazou A, 2009. Regulation of Bcl-2 phosphorylation in response to oxidative stress in cardiac myocytes.*Free Radic Res* 43: 809-816.

Meares GP, Zmijewska AA, Jope RS, 2004. Heat shock protein-90 dampens and directs signaling stimulated by insulin-like growth factor-1 and insulin. *FEBS Lett* 574: 181-186.

Moodie SA, Wolfman A, 1994. The 3Rs of life: Ras, Raf, and growth regulation. Trends Genet 10: 44-48.

Morimoto RI, Sarge KD, Abravaya K. 1992. Transcriptional regulation of heat shock genes. A paradigm for inducible genomic responses. J Biol Chem 267:21987–21990.

Morimoto RI, Santoro MG (1998) Stress inducible responses to heat shock proteins: new pharmacological targets for cytoprotection. Nature Biotechnol 16:833–838.

Neckers L, 2002. Hsp90 inhibitors as novel cancer therapeutic agents.*Trends Mol Med* 8: 555-561.

New L, Han J, 1998. The p38 MAP kinase pathway and its biologicsl function. Trends. Cardicvasc.Med8, 220-229.

Padmini E, Usha Rani M, 2009a. Seasonal influence on heat shock protein 90α and heat shock factor 1 expression during oxidative stress in fish hepatocytes from polluted estuary. *J Exp Mar Biol Ecol* 372: 1-8.

Padmini E, Usha Rani M, 2009b. Evaluation of oxidative stress biomarkers in hepatocytes of grey mullet inhabiting natural and polluted estuaries.*Sci Total Environ* 407: 4533-4541.

Padmini E, Usha Rani M, 2010. Thioredoxin and HSP90α modulate ASK1–JNK1/2 signaling in stressed hepatocytes of Mugil cephalus *Comp Biochem Physiol. Part C* 151187–193

Papp E, Nardai G, Soti Cs, Csermely P, 2003. Molecular chaperones, stress proteins and redox homeostasis. *Biofactors* 17: 249-257.

Park DS, Stefanis L, Yan CYI, Farinelli SE, Greene LA, 1996. Ordering the cell death pathway: differential effects of Bcl-2, an interleukin-1-converting enzyme family protease inhibitor, and other survival agents on JNK activation in serum/nerve growth factor-deprived PC12 cells. *J Biol Chem*271: 21898-21905.

Park J, Kim I, Oh YJ, Lee Kw, Han PL, Choi EJ, 1997. Activation of c-JunN-terminal kinase antagonizes an anti-apoptotic action of Bcl-2. *J BiolChem* 272: 16725-16728.

Parsell DA, Taulien J, Lindquist S, 1993. The role of heat-shock proteins in thermotolerance. *Philos Trans R Soc Lond BBiol Sci*339: 279-285.

Piatelli MJ, Doughty C, Chiles TC, 2002. Requirement for a hsp90 chaperone-dependent MEK1/2-ERK pathway for B cell antigen receptor-induced cyclin D2 expression in mature B lymphocytes. *J Biol Chem* 277: 12144-12150.

Pugazhenthi S, Nesterova A, Sable C, Heidenreich KA, Boxer LM, Heasley LE, Reusch JE, 2000. Akt/protein kinase B up-regulates Bcl-2 expression through cAMP-response element-binding protein. *J Biol Chem*275: 10761-10766.

Rawlings JS, Rosler KM, Harrison DA, 2004. The JAK/STAT signaling pathway. *J Cell Sci* 117: 1281-1283.

Robinson MJ, Cobb MH, 1997. Mitogen-activated protein kinase pathways.*Curr Opin Cell Biol* 9: 180-186.

Sabapathy K, Hochedlinger K, Nam SY, Bauer A, Karin M, Wagner EF, 2004. Distinct roles for JNK1 and JNK2 in regulating JNK activity andc-Jun-dependent cell proliferation. *Mol Cell* 15: 713-725.

Sabapathy K, Hu YL, Kallunki T, Schreiber M, David JP, Jochum W,Wagner EF, Karin M, 1999a. JNK2 is required for efficient T-cell activation and apoptosis but not for normal lymphocyte development. *Curr Biol*11: 116-125.

Sabapathy K, Hu YL, Kallunki T, Schreiber M, David JP, Jochum W,Wagner EF, Karin M, 1999a. JNK2 is required for efficient T-cell activation and apoptosis but not for normal lymphocyte development. *Curr Biol*11: 116-125.

Sabapathy K, Jochum W, Hochedlinger K, Chang L, Karin M, Wagner EF, 1999b. Defective neural tube morphogenesis and altered apoptosis in the absence of both JNK1 and JNK2. *Mech Dev* 89: 115-124.

Sakon S, Xue X, Takekawa M, Sasazuki T, Okazaki T, Kojima Y, Piao JH, Yagita H, Okumura K, Doi T, Nakano H, 2003.NF-κB inhibits TNF-induced accumulation of ROS that mediate prolonged MAPK activation and necrotic cell death. *EMBO J*22: 3898-3909.

Sato S, Fujita N, Tsuruo T, 2000. Modulation of Akt kinase activity by binding to HSP90. *Proc Natl Acad Sci USA* 97: 10832-10837.

Schoof N, Bonin Fv, Trumper L, Kube D, 2009. HSP90 is essential forJak-STAT signaling in classical Hodgkin Lymphoma cells. *Cell Commun Signal* 7: 17-21.

Schouten GJ, Vertegaal ACO, Whiteside ST, Israel A, Toebes M, Dorsman JC, van der Eb AJ, Zantema A, 1997. IκBα is a target for the mitogen-activated 90 kDa ribosomal S6 kinase. *EMBO J*16: 3133-3144.

Schulte TW, Blagosklonny MV, Ingui C, Neckers L, 1995. Disruption of the Raf-1-Hsp90 molecular complex results in destabilization of Raf-1 and loss of Raf-1-Ras association. *J Biol Chem* 270: 24585-24588.

Shang L, Tomasi TB, 2006. The heat shock protein 90-CDC37 chaperone complex is required for signaling by types I and II interferons.*J Biol Chem* 281: 1876-1884.

Sharrocks AD, 2001. The ETS-domain transcription factor family. *Nat Rev Mol Cell Biol* 2: 827-837.

Sharrocks AD, 2002. Complexities in ETS-domain transcription factor function and regulation: lessons from the TCF (ternary complex factor) subfamily. The Colworth Medal Lecture. *Biochem Soc Trans* 30: 1-9.

Shaulian E, Karin M, 2002. AP-1 as a regulator of cell life and death. *Nat Cell Biol* 4: E131-E136.

Shen H-M, Liu Z-g, 2006. JNK signaling pathway is a key modulator in cell death mediated by reactive oxygen and nitrogen species. *Free Radic Biol Med* 40: 928-939.

Shen S, ZhangP, Lovchik MA, Li Y, Tang L, Chen Z, Zeng R, Ma D, Yuan J, Yu Q, 2009. Cyclodepsipeptide toxin promotes the degradation of Hsp90 client proteins through chaperone-mediated autophagy. *J Cell Biol*185: 629-639.

Shen Y-H, Godlewski J, Zhu J, Sathyanarayana P, Leaner V, Birrer MJ, Rana A, Tzivion G, 2003. Cross-talk between JNK/SAPK and ERK/MAPK pathways: sustained activation of JNK blocks ERK activation by mitogenic factors. *J Biol Chem* 278: 26715-26721.

Song G, Ouyang G, Bao S, 2005. The activation of Akt/PKB signaling pathway and cell survival. *J Cell Mol Med*9: 59-71.

Song J, Takeda M, Morimoto RI, 2001. Bag1-Hsp70 mediates a physiological stress signalling pathway that regulates Raf-1/ERK and cell growth.*Nat Cell Biol* 3: 276-282.

Srivastava RK, Mi Q, Hardwick JM, Longo DL, 1999. Deletion of the loop region of Bcl-2 completely blocks paclitaxel-induced apoptosis. Proc Natl Acad Sci U S A. 1999 March 30; 96(7): 3775–3780.

Tanaka, N., Kamanaka, M., Enslen, H., Dong, C., Wysk, M., Davis, R. J., Flavell, R. A. (2002) Differential involvement of p38 mitogen-activated protein kinase kinases MKK3 and MKK6 in T-cell apoptosis. *EMBO Rep.***3**,785-791

Tang F, Tang G, Xiang J, Dai Q, Rosner MR, Lin A, 2002. Absence of NF-κB-mediated inhibition of c-Jun N-terminal kinase activation contributes to tumor necrosis factor alpha-induced apoptosis. *Mol Cell Biol* 24: 8571-8579.

Tang F, Tang G, Xiang J, Dai Q, Rosner MR, Lin A, 2002. Absence of NF-κB-mediated inhibition of c-Jun N-terminal kinase activation contributes to tumor necrosis factor alpha-induced apoptosis. *Mol Cell Biol* 24: 8571-8579.

Tang G, Minemoto Y, Dibling B, Purcell NH, Li Z, Karin M, Lin A, 2001. Inhibition of JNK activation through NF-κB target genes. *Nature*414: 313-317.

Tang G, Minemoto Y, Dibling B, Purcell NH, Li Z, Karin M, Lin A, 2001. Inhibition of JNK activation through NF-κB target genes. *Nature*414: 313-317.

Terry J, Lubieniecka JM, Kwan W, Liu S, Nielsen TO. Hsp90 Inhibitor 17-Allylamino-17-Demethoxygeldanamycin Prevents Synovial Sarcoma Proliferation via Apoptosis in in vitro Models. *Clin Cancer Res.* 2005;**11**:5631–5638

Tournier C, Hess P, Yang DD, Xu J, Turner TK, Nimnual A, Bar-Sagi D,Jones SN, Flavell RA, Davis RJ, 2000. Requirement of JNK for stress-induced activation of the cytochrome c-mediated death pathway. *Science* 288: 870-874.

Verheij M, Bose R, Lin XH, Yao B, Jarvis WD, Grant S, Birrer MJ, Szabo E, Zon LI, Kyriakis JM, Haimovitz-Friedman A, Fuks Z, Kolesnick RN, 1996. Requirement for ceramide-initiated SAPK/JNK signalling in stress-induced apoptosis. *Nature* 380: 75-79.

Verheij M, Bose R, Lin XH, Yao B, Jarvis WD, Grant S, Birrer MJ, Szabo E, Zon LI, Kyriakis JM, Haimovitz-Friedman A, Fuks Z, Kolesnick RN, 1996. Requirement for ceramide-initiated SAPK/JNK signalling in stress-induced apoptosis. *Nature* 380: 75-79.

Wang HG, Takayama S, Rappt UR, Reed JC, 1996. Bcl-2 interacting protein, BAG-1, binds to and activates the kinase Raf-1 *Proc. Natl. Acad. Sci. Cell Biology* 93: 7063-7068.

Wang X, Martindale JL, Liu Y, Holbrook NJ, 1998. The cellular response to oxidative stress: influences of mitogen-activated protein kinase signaling pathways on cell survival. *Biochem J* 333: 291-300.

Wang X, McCullough KD, Franke TF, Holbrook NJ, 2000. Epidermal growth factor receptor-dependent Akt activation by oxidative stress enhances cell survival. *J Biol Chem* 275: 14624-14631.

Wang Z, Wang DZ, Hockemeyer D, McAnally J, Nordheim A, Olson EN, 2004. Myocardin and ternary complex factors compete for SRF to control smooth muscle gene expression. *Nature*428: 185-189.

Waskiewicz AJ, Cooper JA, 1995. Mitogen and stress response pathways: MAP kinase cascades and phosphatase regulation in mammals and yeast.*Curr Opin Cell Biol* 7: 798-805.

Waskiewicz AJ, Cooper JA, 1995. Mitogen and stress response pathways: MAP kinase cascades and phosphatase regulation in mammals and yeast.*Curr Opin Cell Biol* 7: 798-805.

Weston CR, Davis RJ, 2002. The JNK signal transduction pathway. *Curr Opin Genet Dev*12: 14-21.

Widegren, U., Wretman, C., Lionikas, A., Hedin, G. and Henriksson, J. (2000). Influence of exercise intensity on ERK/MAP kinase signalling in human skeletal muscle. *Pflugers Arch.* 441, 317-322.

Yang J, Liu X, Bhalla K, Kim CN, Ibrado AM, Cai J, Peng T-I, Jones DP,Wang X, 1997. Prevention of apoptosis by Bcl-2: release of cytochrome c from mitochondria blocked. *Science* 275: 1129-1132.

Yang X, Khosravi-Far R, Chang HY, Baltimore D, 1997. Daxx, a novelFas-binding protein that activates JNK and apoptosis. *Cell* 89: 1067-1076.

Zhang R, Al-Lamki R, Bai L, Streb JW, Miano JM, Bradley J, Min W, 2004a. Thioredoxin-2 inhibits mitochondria-located ASK1-mediated apoptosis in a JNK-independent manner. *Circ Res* 94: 1483-1491.

Zhang R, Luo D, Miao R, Bai L, Ge Q, Sessa WC, Min W, 2005. HSP90-Akt phosphorylates ASK1 and inhibits ASK1-mediated apoptosis. *Oncogene* 24: 3954-3963.

Zhang S, Lin Z-N, Yang C-F, Shi X, Ong C-N, Shen H-M, 2004b. Suppressed NF-κB and sustained JNK activation contribute to the sensitization effect of parthenolide to TNF-α-induced apoptosis in human cancer cells. *Carcinogenesis* 25: 2191-2199.

INDEX

#

20th century, 133

A

access, 153
accounting, 117
acetylation, 69
acid, 4, 6, 25, 29, 32, 33, 56, 57, 78, 145, 161, 180, 186
activity level, 51, 55
adaptability, 162
adaptation, viii, ix, 2, 15, 31, 34, 51, 54, 55, 56, 58, 62, 74, 77, 92, 110, 122, 175, 180, 181
adaptations, 34, 54, 56
adenine, 62
adjustment, 136
adults, 90, 153, 157, 161
aerobic organisms, viii, 31
Africa, 125
age, vii, 1, 2, 59, 82, 91, 118, 159, 160, 162, 163, 164, 168, 170
agencies, 124, 126
agriculture, 15, 124
air temperature, 91
Alaska, 171
albumin, 36
allele, 95
allometry, 77
alters, 123
amino, 18, 29, 74, 180
amino acid, 18, 29, 74, 180
amphibians, 68, 73
anatomy, 136
ancestors, 109
anesthetics, 134
Anguilla, ix, 56, 99, 100, 101, 102, 109, 110, 111, 112, 113
ANOVA, 6
anoxia, 3, 26, 27
anthropogenic factors, vii, 2
anti-apoptotic role, 174
antigen, 189
antioxidant, vii, viii, 2, 23, 26, 27, 31, 32, 33, 34, 35, 41, 44, 48, 49, 50, 51, 52, 53, 54, 55, 56, 58, 59, 180
apoptosis, xi, 32, 173, 176, 177, 178, 179, 180, 181, 182, 183, 184, 185, 186, 187, 188, 190, 191, 192
apoptosome formation, xi, 173
apoptotic events, xi, 173
aquaculture, x, 3, 20, 34, 57, 94, 115, 126, 153
aquatic systems, 122, 124, 128
Argentina, vii, 131, 132, 148, 149
aromatic hydrocarbons, 174
arsenic, 174
artemia, 22, 23, 25, 28
arthropods, 77
ascorbic acid, 57
Asia, 155, 165
assessment, 24, 34, 165
autooxidation, 33
avoidance, 110, 145

B

backwaters, 158
bacteria, viii, 61, 62, 63, 73, 74, 75, 76, 174
Bangladesh, 152, 153, 156, 170
barriers, 153
base, viii, ix, 61, 62, 63, 64, 66, 68, 72, 73, 74, 75, 76, 78, 132
base pair, 62, 72
Beagle Channel, vii, x, 131, 132, 133, 134, 135, 136, 137, 138, 139, 140, 144, 145, 146, 147, 148, 149

behaviors, x, 99, 108
Belgium, 154, 163
benefits, 117
Bengal, Bay of, 152, 170
benthic invertebrates, 27
bias, ix, 62, 63, 69, 72, 73, 76, 85
biased gene conversion (BGC), ix, 62
bioaccumulation, 21, 26
biochemistry, 21, 57, 129
biodiversity, 56, 116, 117, 127, 128
biogeography, 166
bioinformatics, 75
biological processes, vii, 1, 2, 66, 180
biological systems, 32
biomarkers, 2, 25, 55, 58, 189
biomass, 134, 144, 160
biotic, vii, ix, 24, 32, 55, 57, 81, 82, 92, 170
biotic factor, ix
birds, 34, 68
bisphenol, 21
Black Sea, vii, viii, 1, 3, 4, 7, 8, 10, 16, 19, 25, 26, 27, 28, 31, 34, 35, 36, 37, 38, 39, 40, 41, 42, 43, 45, 46, 47, 48, 50, 51, 55, 56, 58, 158, 166
Black Sea fish species, vii, 1, 3, 8, 27, 35, 37, 38, 39, 40, 42, 45, 47, 48, 50, 55, 56, 58
Black Sea region, 58
blastula, 16
blood, 21, 34, 35, 36, 38, 39, 40, 41, 49, 51, 52, 53, 54, 58, 175
body size, 64, 66, 82, 88, 97
body weight, 163
bonds, 32, 33
bone, 51, 112, 132, 136, 137, 140, 142
bone form, 51
bones, 132, 137, 145
brain, 175
Brazil, x, 56, 115, 116, 117, 118, 119, 120, 121, 122, 123, 125, 126, 127, 128, 129, 185
breakdown, 112
breeding, 83, 88, 91, 97, 100, 101, 108, 109, 127
brevis, 121, 123, 124, 125, 126
Britain, 169, 170
browser, 70
Burma, 153

C

cadmium, 174
calcium, 100, 112, 113
calibration, 170
calorimetric measurements, 6
calorimetric method, 25
calorimetry, 2, 27
cancer, 189, 192
cancer cells, 192
canonical correspondence analysis, 160
capillary, 184, 186
carbohydrates, 18, 26
carbon, 123, 128
carotenoids, viii, 31, 32, 33, 37, 41, 44, 46, 49, 50, 57
cartilaginous, 138
case studies, 128
case study, 149
Caspian Sea, 16
catadromous, vii, ix, 99, 100, 101, 107, 109, 112, 113
cell cycle, 180
cell death, 175, 176, 177, 179, 183, 184, 188, 189, 190
cell division, 10, 18, 176
cell fate, 179
cell line, 182
cell membranes, 18, 32, 33
cell signaling, 185
cellular homeostasis, 174, 180, 182
central nervous system, 183
challenges, x, 58, 116
changing environment, 174
chaperones, 174, 179, 186, 189
characiformes, x, 115, 120, 121, 122, 123
chemical, vii, 2, 15, 18, 20, 21, 23, 56, 74, 90, 154
chemical characteristics, 90
chemical structures, 23
chemicals, 14, 21, 23, 55, 134
Chile, 148
chimpanzee, 63, 73
China, xi, 152, 153, 155, 163
chlorine, 174
chloroform, 4
choline, 18
chromatograms, 4
chromatography, 4, 26
chromium, 174
classes, 26, 27, 44, 64, 76, 159
classification, ix, 72, 99, 107, 152
climate, 24, 28, 117, 128, 153, 154
climate change, 24, 153, 154
climatic factors, 118, 164, 167
clinical application, 186
clustering, 88, 91
clusters, 78, 86
coding, 62, 63, 68, 69, 72, 73, 76
codon, 62, 63, 79
collaboration, 126
colonization, 90, 91, 93, 134

commercial, xi, 57, 100, 125, 151, 153
communication, 133
communities, 94, 116, 117, 123, 124, 127, 132
community, 125
compaction, 142
competition, x, 63, 76, 91, 97, 99, 107, 109, 162
competitors, 160
complexity, 69, 70, 149
composition, vii, viii, ix, x, 1, 2, 3, 4, 6, 7, 8, 9, 16, 18, 20, 24, 25, 58, 61, 62, 63, 64, 66, 68, 72, 73, 74, 75, 76, 78, 86, 91, 92, 94, 111, 116, 128, 132, 134, 142, 144, 147, 148, 160, 176
compositional structure, 63
compounds, viii, 4, 8, 12, 15, 21, 31, 32, 33, 37, 51, 52, 53, 103
computer, 93
conception, 166
conduction, vii, 1, 4
conference, 168
conjugation, 52
conservation, viii, ix, 32, 73, 75, 81, 82, 83, 92, 93, 94, 109, 116, 126, 127, 128, 131, 132, 149, 155, 166, 168
consolidation, 134
construction, x, 91, 94, 115, 126, 153
consumption, vii, 1, 2, 10, 16, 18, 20, 32, 51, 54, 55, 136, 145, 163
containers, 35
contaminant, 54, 55, 184
contamination, viii, 2, 24, 31, 164
continental, 165
contradiction, 73
control group, 5
copper, 174
correlation, viii, 6, 8, 20, 21, 22, 38, 49, 55, 61, 63, 66, 68, 69, 71, 73, 74, 76, 86, 144, 175, 182
correlation coefficient, 6, 8, 38, 66, 69, 71
correlations, 44, 48, 50, 54, 69, 71
cost, 26, 88
critical period, 3, 20, 170
cultivation, 124
culture, x, 26, 110, 117, 118, 124, 125, 136, 155, 166
cyanide, 174
cycles, 112, 118, 132
cycling, 33, 174, 175
cytochrome, 191, 192
cytokines, 176, 178, 179
cytometry, 79
cytoplasm, 184
cytosine, 62, 67, 68, 72, 76

D

damages, 117, 125
danger, ix, 81, 92
Danube River, 158
data set, 66, 70
database, 79
defense mechanisms, viii, 31, 34, 51, 53, 54
deficiency, 57
deforestation, x, 115, 124
degradation, ix, x, 33, 81, 92, 116, 117, 118, 123, 124, 126, 182, 184, 186, 187, 188, 190
deoxyribonucleic acid, 74, 78
dephosphorylation, 188
deposition, 132, 164
deprivation, 175
depth, 136, 145, 157, 158, 160
destruction, 126, 180
detachment, 179
detectable, 32, 49, 50
detection, 82
detergents, 175
detoxification, 34, 52, 54
developmental change, 24
diatoms, 160
dichotomy, 89
diet, 55, 116, 136, 160, 163, 164, 169, 170
diet composition, 164
direct measure, 21
discrimination, 164
diseases, 2
dissociation, 188
dissolved oxygen, 123, 154, 163
distilled water, 35, 102
distribution, ix, x, 62, 69, 72, 73, 78, 79, 82, 83, 85, 86, 90, 91, 93, 95, 96, 107, 109, 111, 118, 122, 131, 132, 134, 136, 144, 145, 146, 147, 154, 155, 158, 160, 161, 164, 167, 169, 179
divergence, 82, 88, 89, 96, 97
diversity, x, 64, 92, 93, 97, 101, 116, 117, 118, 122, 125, 126, 132, 137, 144, 147
DNA, ix, 24, 32, 62, 63, 64, 68, 69, 70, 72, 73, 75, 76, 77, 78, 79, 81, 83, 86, 88, 90, 92, 93, 94, 95, 96, 97, 110, 111, 161, 170, 171, 175, 187
DNA repair, 63
domestication, xi, 152, 155
dominance, 124
double bonds, 33
down-regulation, 183
drainage, 96, 118, 133, 168
drinking water, 117
drought, 117, 123, 127

E

early warning, 3
Eastern Europe, xi, 151, 166
ecology, vii, viii, xi, 2, 8, 32, 41, 53, 56, 57, 77, 94, 112, 127, 128, 132, 134, 145, 147, 149, 152, 155, 156, 161, 165, 166, 167, 171
economic losses, 124
ecosystem, x, xi, 15, 116, 123, 126, 131, 132, 136, 145, 147
editors, 126, 128, 129
effluents, 126
egg, xi, 4, 5, 6, 8, 9, 10, 16, 17, 19, 23, 152, 156, 157, 169
election, 96
electron, 33, 102, 103
electrons, 32
embryogenesis, vii, 1, 2, 5, 6, 8, 10, 12, 16, 17, 18, 20, 24, 28, 52
emigration, 158, 171
endocrine, 21, 56
endocrinology, 175
endothelial cells, 184, 186
energy, vii, 1, 2, 3, 10, 15, 16, 17, 18, 20, 21, 23, 24, 28, 64, 123, 180
energy consumption, 10, 16
energy expenditure, 21
energy generation, vii, 1, 2, 15, 16, 17, 18, 20, 21
environment, viii, 5, 20, 23, 24, 31, 34, 35, 50, 52, 53, 54, 55, 62, 63, 64, 68, 72, 74, 105, 107, 110, 125, 128, 132, 134, 144, 145, 161, 174, 175, 180
environmental change, 153
environmental conditions, 82, 88, 108, 110, 125, 153, 158, 161
environmental contamination, viii, 24, 31
environmental factors, viii, 2, 20, 24, 31, 32, 63, 73, 156
environmental impact, 117
environmental stimuli, 177
environmental stress, 3, 20, 21, 56, 173, 174, 175, 176, 178, 185
environmental variables, 164
enzymatic activity, 34, 35, 41, 44, 47, 51, 52, 55
enzyme, 21, 23, 26, 27, 32, 35, 36, 38, 40, 41, 43, 44, 46, 48, 49, 50, 51, 52, 53, 54, 55, 57, 58, 189
enzymes, viii, 2, 24, 31, 32, 33, 49, 50, 51, 53, 54, 58
epithelium, 23, 104
equilibrium, 173
equipment, 134
erythrocytes, 49, 57
estrogen, 176
ethanol, 4, 21, 38, 102
ethnographers, 144
eukaryotes, viii, 61, 73
eukaryotic, 179, 180
eukaryotic cell, 180
Europe, xi, 90, 151, 155, 166, 169
evaporation, 117
evapotranspiration, 124
evidence, xi, 52, 63, 97, 101, 108, 132, 144, 146, 151, 152, 154, 158, 161, 168, 177, 183
evolution, viii, ix, 31, 56, 62, 63, 64, 69, 73, 75, 76, 78, 94, 101, 111, 113, 166
excision, 78
exercise, 175, 176, 192
exotic species, x, 115, 121, 125
experimental condition, 23
exploitation, 145, 146, 153, 171
exposure, 2, 3, 15, 20, 21, 23, 24, 56, 123, 162, 175, 180
extinction, 36, 37, 91, 123, 153, 155
extracellular matrix, 103, 179
extracts, 4, 37, 38

F

families, 20, 73, 116, 120, 121, 134, 137, 138, 140, 144, 179, 183
fat, 50, 52, 54
fat soluble, 52
fatty acids, vii, 1, 6, 7, 17, 18, 26, 51
fauna, x, 115, 118, 120, 122, 123, 126, 127, 134, 144, 147, 148, 149
fertilization, 16, 18, 50
fibroblasts, 176
fidelity, 89
Fiji, 148
fish eggs, vii, 1, 3, 4, 6, 14, 20, 23, 25, 28, 169
fish embryogenesis, 1, 7, 8, 18
fisheries, 57, 83, 86, 100, 124, 126, 127, 129, 146, 153, 155, 165, 166
fishing, 86, 118, 126, 128, 132, 136, 144, 146, 149, 153
fitness, 91, 101
floods, 123
fluctuations, vii, viii, 1, 3, 6, 7, 8, 18, 19, 31, 53, 166
food, 12, 16, 18, 32, 53, 55, 91, 101, 108, 110, 125, 127, 136, 145, 162, 164, 170, 171, 175
food chain, 12, 53, 55
food habits, 125
food web, 164
force, 73, 128
Ford, 93
forebrain, 177
formation, xi, 10, 16, 21, 32, 51, 53, 54, 58, 63, 64, 76, 173, 174, 177, 183, 186

formula, 137
fragile site, 73
fragments, 145
France, xi, 108, 152, 154, 155, 157, 160, 162, 163, 164, 165, 166, 167, 168, 169, 170, 171
free radicals, viii, 31, 32, 50, 175
freshwater, vii, ix, x, xi, 27, 82, 93, 95, 96, 97, 99, 100, 101, 102, 103, 105, 106, 107, 108, 109, 110, 111, 112, 113, 115, 116, 117, 118, 122, 123, 124, 125, 126, 127, 128, 151, 152, 153, 154, 156, 158, 160, 161, 162, 169, 170, 171
freshwater eel, vii, ix, 99, 100, 101, 102, 103, 105, 106, 107, 108, 109, 110, 111, 112
funding, 153
fungi, 76

G

gamete, 56, 125, 157
gastrula, 16
GC-content, 76
gene expression, 2, 74, 76, 176, 177, 181, 191
genes, 64, 69, 70, 76, 78, 79, 179, 180, 183, 184, 189, 191
genetic diversity, 92, 97
genetics, 75, 94, 110
genome, viii, ix, 62, 63, 64, 66, 68, 69, 70, 72, 73, 74, 75, 76, 78, 79
genomics, 75
genre, 165
genus, ix, 83, 95, 96, 99, 100, 106, 109, 111, 136, 140, 146, 152
geometry, 77
Georgia, 149
Germany, 4, 154, 155, 157, 160, 162, 163, 164, 169, 170
gill, 23
global climate change, 24
glutathione, 32, 33, 34, 36, 37, 40, 41, 49, 50, 51, 52, 53, 54, 57
glycine, 178
glycogen, 18
glycosaminoglycans, 103
gonads, 27, 35, 45, 46, 47, 48, 49, 50, 51, 52, 53, 100
Greece, 152
grouping, 74, 145
growth, vii, ix, 1, 2, 16, 18, 20, 23, 25, 26, 52, 63, 76, 94, 95, 96, 99, 100, 101, 103, 107, 108, 109, 124, 125, 146, 160, 161, 162, 163, 164, 167, 168, 170, 171, 176, 177, 178, 180, 182, 183, 184, 189, 190, 191
growth factor, 176, 177, 178, 182, 183, 189, 191
growth rate, 160, 161, 163
growth temperature, 63, 76
guanine, 62, 72, 76
Gulf of Mexico, 168

H

habitat, viii, ix, 61, 65, 66, 67, 68, 72, 73, 74, 81, 83, 91, 92, 94, 95, 96, 97, 99, 100, 101, 103, 106, 107, 108, 109, 110, 111, 112, 113, 118, 156, 157, 160, 163, 165, 169, 170
habitats, viii, ix, x, xi, 2, 16, 54, 61, 62, 63, 64, 67, 68, 72, 73, 74, 99, 100, 101, 103, 105, 106, 107, 108, 109, 110, 116, 124, 151, 152, 168
haplotypes, 84, 85, 93
health, 56, 57, 174
heat dissipation, vii, 1, 4, 5, 8, 9, 10, 11, 15, 19, 21, 23
heat shock protein, 174, 179, 180, 185, 186, 189, 190
Heat shock proteins, xi, 173, 174, 188
heavy metals, 15, 174, 175, 180
height, 16
hemoglobin, 56
hepatocytes, 179, 180, 181, 183, 184, 185, 189
heterochromatin, 78
heterogeneity, 63, 73, 75, 78
hexane, 4, 5, 12, 13, 14, 15
histone, 69
history, xi, 64, 82, 83, 88, 90, 93, 95, 96, 97, 100, 101, 102, 103, 105, 106, 110, 111, 112, 113, 127, 147, 152, 155, 156, 157, 160, 164, 165, 168, 170
holocene, 132, 149
homeostasis, 64, 174, 176, 178, 180, 181, 182, 189
homologous chaperone proteins, xi, 173
house, 127
human, ix, xi, 12, 21, 36, 57, 62, 63, 64, 73, 75, 76, 78, 79, 81, 86, 92, 93, 116, 122, 124, 131, 132, 154, 175, 192
human activity, 12, 86, 124
human genome, 63, 75, 78, 79
human health, 57
hunter-gatherers, xi, 131, 136, 146
hybridization, xi, 151, 152, 165
hydrocarbons, 12
hydrogen, 32, 33, 64, 175
hydrogen bonds, 64
hydrogen peroxide, 32, 33, 175
hydroperoxides, 32, 33, 49, 52
hydrophobicity, 64
hydroquinone, 33
hydroxyl, 32, 175
hyperthermia, 174
hypothesis, 63, 64, 73, 97, 109, 113, 156, 162, 170
hypoxia, 3, 27, 174, 175

I

IBD, 86, 88
Ichthyofauna, x, 115
ideal, 125
identification, ix, 20, 57, 70, 81, 82, 93, 126, 136, 144
immune response, 178
immune system, viii, 31, 185
imprinting, 90
in vitro, 174, 191
in vivo, 2
India, xi, 151, 153, 160, 163, 167, 173
individual character, 18
individuals, 4, 20, 83, 85, 86, 88, 90, 101, 132, 134, 136, 138, 139, 145, 156, 158, 162
Indonesia, ix, 99, 100, 102, 107, 108, 111
induction, 54, 55, 175, 178, 180, 183, 185, 188
industries, 117
infection, 2, 163, 175, 180
inflammation, 188
inhibition, 36, 179, 181, 182, 183, 184, 185, 188, 191
inhibitor, 184, 188, 189
injury, 187
inner ear, 104
insects, 160, 163
insertion, 79
institutions, 154
insulin, 178, 189
integrity, x, 115, 125, 142
interference, 123
interferons, 190
intestine, 16, 18
introns, 62, 69, 70, 79
invertebrates, 12, 19, 21, 23, 25, 27, 33, 35, 51, 54, 145, 146
investment, 146
iodine, 37
ionizing radiation, 178
Iowa, 168
Iran, 155
Iraq, 155
Ireland, xi, 102, 151, 152
iron, 174
irradiation, 24
irrigation, 117, 124
ischemia, 178, 186
islands, 78
isolation, 82, 91
isopods, 136, 160
isotope, 165, 166
isozymes, 177
Israel, 190
issues, 27, 41, 55, 83, 91, 144
Italy, 61, 93, 152

J

Japan, 81, 82, 83, 85, 90, 91, 94, 95, 96, 99, 102, 108, 109, 110, 111, 112
Japan, Sea of, 83, 85, 91
Java, 107
jumping, 153
juveniles, xi, 106, 152, 153, 155, 161, 162, 163, 164, 172

K

keratinocyte, 186
keratinocytes, 178
kidney, 175
kill, 21, 162
kinase activity, 190
KOH, 38
Korea, 83, 90

L

lakes, xi, 100, 118, 124, 128, 151, 152, 153, 160
landings, 123
landscape, 132
landscapes, 75
larva, 27, 106
larvae, vii, xi, 1, 2, 3, 4, 5, 6, 7, 8, 9, 10, 11, 12, 13, 14, 15, 16, 17, 18, 19, 20, 21, 22, 23, 24, 25, 26, 27, 28, 29, 58, 100, 152, 153, 155, 158, 159, 160, 161, 162, 163, 164, 167, 169, 170, 171
larvae hatching, vii, 1, 2, 4, 17
larval development, 57
larval stages, 153
lead, 86, 91, 93, 123, 124, 132, 153, 174, 177
leisure, 123
life cycle, 27, 53, 92, 100, 107, 112, 123
light, ix, 62, 64, 73, 102, 158, 177
linear model, 163
lipid AOA, 45, 48, 52, 53
lipid composition, vii, 1, 3, 4, 6, 7, 8, 18, 20, 58
lipid metabolism, 2, 23, 29
lipid peroxidation, viii, 2, 23, 26, 27, 31, 33, 52, 58
lipid peroxides, 52
lipids, vii, 1, 2, 6, 8, 16, 18, 24, 26, 32, 38, 41, 43, 44, 49, 50
liquid chromatography, 62
liver, 16, 27, 35, 36, 44, 45, 46, 50, 51, 52, 53, 54, 55, 58, 181, 182, 186

local conditions, 91
localization, 186
loci, 86, 90
logging, 124
Luo, 192
lymphocytes, 189
lymphoma, 179, 182

M

machinery, 186
macromolecules, 175
magnitude, 63, 123
majority, 46, 50, 53, 73, 90, 162
mammals, 32, 34, 63, 68, 73, 145, 174, 191
management, x, 92, 95, 100, 113, 116, 126, 128, 132, 148, 149, 165
manganese, 174
MAPK/ERK, 176
mapping, 73, 186
marine environment, 76, 101, 109, 145
marine fish, vii, viii, x, 27, 29, 31, 57, 58, 131, 132, 137, 161, 171
marine species, 109
Maryland, 168
masking, 70
mass, viii, 25, 61, 62, 64, 66, 72, 77, 118, 122
masu salmon, vii, ix, 81, 82, 83, 85, 86, 87, 88, 90, 91, 92, 93, 94, 95, 96, 97
matrix, 103, 179
matter, viii, 61, 72
measurement, 57
measurements, 2, 4, 5, 6, 21, 26, 28, 36, 103, 118
median, 69, 158
Mediterranean, 97, 109, 156, 164
MEK, 176, 182, 187
melting, 133
membranes, viii, 18, 31, 32, 33
memory, 153
mercury, 25, 28, 164, 174
messengers, 175
metabolic, vii, 1, 4, 8, 12, 13, 14, 15, 17, 18, 19, 20, 21, 22, 24, 25, 26, 27, 28, 64, 72, 77
metabolic pathways, 15, 24, 32, 52, 175
metabolism, 2, 16, 18, 21, 23, 24, 26, 27, 29, 34, 53, 54, 57, 58, 64, 66, 72, 77, 123, 176, 180, 186
metabolites, 15, 18, 52
metals, 15, 23, 174, 175, 180
metamorphosis, 25, 103, 110, 112
metaphase, 79
meter, 136
methodology, 156, 161
methyl group, 78
methylation, viii, 61, 62, 67, 68, 69, 73, 74, 76, 78
Mexico, 117, 168
microbial communities, 74
microcalorimetry, vii, 1, 2, 3, 5, 19, 20, 24, 25, 26, 28
microenvironments, 134
microgeographic scale, ix, 81, 82, 83, 86, 90, 92
microorganisms, 4
microsatellites, 166
microscope, 102
microstructure, 110, 112
Middle East, xi, 152, 155, 167
migration, ix, 82, 83, 88, 89, 90, 91, 94, 99, 100, 101, 106, 107, 108, 109, 110, 111, 112, 152, 154, 156, 161, 162, 164, 165, 167, 168, 171, 176, 178
mitochondria, 32, 49, 57, 181, 182, 192
mitochondrial DNA, ix, 81, 83, 93, 95, 96, 97
mitogen, 183, 185, 187, 188, 190, 191
mitogens, 176
model system, 20
models, 160, 161, 163, 174, 178
modifications, 7, 8, 18, 20, 21, 36, 38
molecular oxygen, 32
molecular weight, viii, 2, 31, 32, 33, 34, 42, 47, 49, 50, 51, 52
molecules, viii, 12, 31, 178, 181, 182
mollusks, 54
moratorium, 153
Morocco, 152, 154, 160, 163, 164
morphogenesis, vii, 1, 2, 16, 18, 24, 190
morphology, 92
mortality, 2, 127, 153, 159, 160, 162, 164, 167, 168, 170
mortality rate, 159, 160
mosaic, 75
Moscow, 25, 26, 56, 57, 58
motif, 176, 186
motivation, 153
mtDNA, 83, 84, 85, 90, 93, 110, 111, 166
multi-protein complexes, 180
muscles, 35, 43
mutation, 63, 68, 75, 76, 97
mutations, 63, 75, 94
myoblasts, 178

N

NAD, 33
NADH, 36, 85, 95
native population, 83
native species, x, 115, 116, 117, 121, 123, 125
natural selection, 63
necrosis, 179, 187, 188, 191

negative consequences, 56
negative effects, 23
negative influences, 92
neotropical region, x, 115, 117, 126
nerve, 16, 183, 189
nerve growth factor, 183, 189
nervous system, 16, 183
Netherlands, 58, 163
neuroglioma, 183
neuronal cells, 177
neurons, 183
neutral, vii, 1, 2, 16, 18, 63, 72, 74, 75
neutral lipids, vii, 1, 2, 16, 18
New Zealand, 102, 111
next generation, 109, 110
nickel, 27, 174
Nile, 121, 125
nitric oxide, 175, 186
nitric oxide synthase, 186
nitrogen, 32, 124, 190
non-enzymatic antioxidants, 44, 51, 52, 54
North America, xi, 90, 112, 151, 152, 153, 155, 164, 165, 166, 171
North Sea, 109, 171
Norway, 112, 154
Nototheniidae family, x, 131, 140, 144
nuclear genome, 79
nucleic acid, 161
nucleosome, 64, 76
nucleotides, 72
nucleus, 49, 176, 178, 184
null, 75
null hypothesis, 75
nutrient, 124
nutrients, 57
nutrition, 54, 55, 57, 116
nutritional status, 18

O

obstacles, 153
oceans, 133
oil, 16
olfaction, 166
operations, 124, 134
opportunities, 77, 166
organ, 50, 51, 54
organic compounds, 12, 15
organism, viii, 2, 17, 20, 24, 28, 31, 32, 33, 34, 49, 51, 52, 53, 56, 63, 64, 69, 72
organism changes, 2
organs, 10, 34, 35, 161, 175
ovaries, 52
overlap, 73
overproduction, 175
overweight, 142
oxidation, 10, 18, 25, 38, 49, 51, 175
oxidative damage, 33, 57
oxidative stress, viii, 31, 32, 33, 34, 49, 50, 51, 52, 53, 55, 56, 174, 175, 178, 179, 180, 181, 182, 183, 184, 185, 187, 188, 189, 191
oxygen, vii, viii, 1, 2, 16, 20, 23, 24, 31, 32, 50, 51, 53, 54, 57, 63, 64, 66, 73, 123, 154, 163, 175, 190
oxygen consumption, vii, 1, 2, 20, 32, 51, 54, 64, 66, 73
oxyradical flux, viii, 31
oysters, 25

P

Pacific, 83, 85, 93, 94, 96, 100, 109, 133
paclitaxel, 191
Panama, 125
parallel, 157
parasites, 163, 164, 170
parentage, 26
parental care, 125
pathology, 2, 57
pathways, xii, 15, 23, 32, 52, 78, 153, 173, 174, 175, 176, 177, 178, 179, 180, 182, 183, 184, 185, 186, 187, 188, 189, 190, 191
PCP, 21, 22, 26
peptide, 180, 181
perciformes, x, 67, 115, 120, 121, 122
periodicity, 160, 164
permit, 134, 144
peroxidation, viii, 2, 23, 26, 27, 31, 33, 52, 58
peroxide, 32, 33, 175
Persian Gulf, xi, 151, 153
pesticide, 5, 6, 15, 23, 24
phagocyte, 56
phenol, 21, 57
phenotype, 95
Philippines, 107, 111
phosphate, 36
phosphoinositides, 179
phospholipids, 6, 8, 17
phosphorus, 4, 15
phosphorylation, 21, 176, 177, 178, 179, 181, 182, 183, 184, 185, 186, 187, 189
photographs, 102
photoresponse, 158, 164
phylogenetic tree, 96
physico-chemical parameters, 74
physiological, 3, 25, 27, 28, 128
physiological factors, 101

physiology, 21, 56, 64, 162, 175
phytoplankton, 127
pigmentation, 16
plankton, 37, 55, 160
plants, 58, 77
plasma membrane, 179
plasticity, xi, 106, 109, 112, 125, 126, 127, 151, 152
playing, xi, 173, 174
PMS, 36
Poland, 28
polar, viii, 61, 66, 68
pollutants, vii, x, 1, 2, 12, 16, 20, 23, 26, 27, 28, 53, 55, 56, 116, 126, 174, 185
pollution, vii, viii, xi, 1, 2, 3, 15, 22, 24, 25, 31, 32, 34, 53, 54, 55, 56, 58, 151, 153, 175, 180
polychlorinated biphenyl, 174
polycyclic aromatic hydrocarbon, 174
polymorphism, 62, 63, 82, 83, 86, 97
polymorphisms, 92
polypeptides, 178
polyunsaturated fat, 51
polyunsaturated fatty acids, 51
population, ix, x, 24, 74, 76, 81, 82, 83, 85, 86, 88, 90, 91, 92, 94, 96, 97, 110, 115, 116, 117, 124, 153, 155, 163, 166, 167, 168, 175
population dynamics, ix, 81, 82, 166, 167
population growth, 94
population size, 76, 86, 91, 92
population structure, 83, 94, 96
Portugal, xi, 151, 152, 153, 154, 157, 159, 160, 161, 166, 170
positive correlation, viii, 61, 68, 74, 182
precipitation, 117, 122, 124
predation, 109
predators, 55, 136, 160
predictability, xi, 151
preparation, 35
preservation, 63, 92, 142, 143
probability, 108, 154
project, 146, 155
prokaryotes, viii, 61, 63, 73, 76
proliferation, x, 115, 125, 177, 178, 180, 182, 183, 184, 189
promoter, 179
prooxidant-antioxidant balance, viii, 31
propagation, 83, 95, 155
prostaglandins, 32
protection, 2, 15, 20, 24, 49, 50, 51, 175, 182
protective mechanisms, 56
protein folding, 174, 180, 188
protein kinase C, 179, 186
protein kinases, 179, 185
protein structure, 64, 76
protein synthesis, 161
proteins, xi, 32, 37, 63, 73, 74, 79, 161, 173, 174, 175, 176, 177, 178, 179, 180, 181, 182, 183, 184, 185, 186, 188, 189, 190
proto-oncogene, 183
purification, 123
pyrimidine, 78
pyrophosphate, 36

Q

quantification, 82
quantity of early life stages, vii, 1, 2
quartile, 69
quinones, 33

R

radiation, 75, 122, 178, 180
radical formation, 58
radicals, viii, 31, 32, 33, 50, 175
radius, 102, 103
rainfall, x, 116, 117, 119, 122, 126, 133, 160, 162
reactions, 33, 66
reactive oxygen, 2, 23, 24, 32, 57, 190
recall, 69, 73
receptors, 176, 178
recombination, 63, 72, 73, 75, 79
recommendations, 57
reconstruction, 165
recovery, xi, 136, 163, 173
recreation, 124
recreational, xi, 100, 151, 153
recruiting, 111
recurrence, 117, 145
recycling, 32
red blood cells, 34, 35, 38, 39, 40, 49, 51
regeneration, 186
regions of the world, 117
regression, 6, 66, 69, 71, 154, 163
regression analysis, 6
regression line, 71
regression model, 154
regulations, x, 86, 116, 126
relevance, 63, 95, 148
repair, 23, 34, 63, 72, 78
reproduction, 23, 53, 56, 100, 109, 111, 126, 169
requirements, 23, 110, 145
researchers, 34, 53, 54, 55, 134
residues, 21, 27
resistance, 24, 34, 56, 125, 175, 182, 185
resolution, 165

resource availability, 20
resources, xi, 16, 63, 76, 100, 117, 132, 134, 144, 146, 151, 152, 171
respiration, 2, 10, 16, 18, 20, 26, 50
response, x, xi, 3, 5, 15, 20, 22, 27, 53, 55, 58, 99, 149, 158, 173, 174, 175, 176, 177, 179, 180, 181, 182, 183, 184, 185, 189, 191
restoration, xi, 128, 152, 153, 155
restoration programs, xi, 152, 153, 155
restriction fragment length polymorphis, 97
restrictions, x, 116, 126
restructuring, 165
reticulum, 32
revenue, x, 116
rhythm, 156
risk, viii, 24, 31, 34, 91
risk assessment, 24, 34
river systems, 91, 153, 155
RNA, 63, 73, 161, 170, 171
Romania, 158
routes, 90
rural population, 116
Russia, 83, 90, 96

S

salinity, vii, 1, 2, 32, 74, 101, 106, 107, 113, 133, 152, 156, 157, 160, 163, 164, 172
salmon, vii, ix, 21, 26, 27, 81, 82, 83, 85, 86, 87, 88, 90, 91, 92, 93, 94, 95, 96, 97, 101, 108, 109, 112, 113
Salmonid species, ix, 81
saltwater, 154
samplings, 134, 136
scaling, 64, 77
scaling law, 64, 77
scavengers, viii, 31, 32, 49, 50, 51, 52
scope, xi, 131, 132
seasonal changes, 170
seasonal flu, viii, 31
seasonality, 117, 118, 125, 127
security, 58
sediment, 27, 35, 124
sedimentation, 37
sediments, 12, 34, 53, 54, 55, 136, 142
selenium, 33, 57, 174
semiarid aquatic ecosystems, x, 116, 123
senility, viii, 31
sensitivity, 21, 56
sensitization, 192
sequencing, 75
serine, 178, 183, 184, 187
serum, 35, 36, 38, 39, 40, 49, 53, 177, 189
serum albumin, 36
severe stress, 184
sewage, 15, 21, 26, 126, 154
sex, 46, 51, 82, 90, 93, 156
sex ratio, 156
Shads, v, xi, 151, 152, 153, 166
shape, 73, 76
Shatt al Arab, 155
shelter, 145
shock, xi, 173, 174, 175, 177, 179, 180, 181, 185, 186, 187, 188, 189, 190
shoreline, 136
shores, 133, 134
shortage, viii, 61, 68
showing, 63, 69, 73, 86, 122, 143
signal transduction, 32, 174, 175, 179, 180, 181, 185, 188, 192
signaling pathway, xii, 173, 176, 178, 179, 180, 182, 183, 184, 185, 186, 187, 188, 189, 190, 191
signalling, 185, 186, 190, 191, 192
signals, 86, 176, 177, 178, 179, 181, 183, 184, 185, 186
significance level, 38
siluriformes, x, 115, 121, 122, 123
silver, 100, 101, 107, 109, 110, 111
skeletal muscle, 176, 192
skeletal remains, 144
smooth muscle, 183, 188, 191
smooth muscle cells, 183, 188
SNP, 62, 63
sodium, 36
software, 72
Solea senegalensis, 26, 27
solution, viii, 31, 35, 37
somatic cell, 73
South America, 117, 122, 123, 127, 128, 147, 148, 149
Southeast Asia, 155, 165
Spain, 152, 154
speciation, 113
species richness, 117
specific tax, 140
spectrophotometric method, 36, 37
spending, ix, 99
spinal cord, 16
Sri Lanka, 170
stability, ix, xi, 63, 73, 94, 131, 132, 177, 180, 186
stabilization, 182
standard error, 142
starvation, 26
state, 15, 82, 118, 124, 175, 178, 181, 184
states, 117, 180
stereomicroscope, 136

sterile, 4, 5, 6
sterols, 6, 7, 8, 18
stimulus, 176, 178
storage, 64, 118, 123
stress, viii, xi, 3, 15, 20, 21, 27, 31, 32, 33, 34, 49, 50, 51, 52, 53, 55, 56, 57, 59, 69, 73, 74, 127, 173, 174, 175, 176, 177, 179, 180, 181, 182, 183, 184, 185, 187, 188, 189, 190, 191
stressors, vii, 2, 3, 20, 21, 174, 175
strontium, 100, 112, 113
structure, vii, ix, x, 7, 8, 20, 64, 69, 79, 81, 82, 83, 85, 86, 88, 90, 91, 92, 93, 94, 95, 96, 97, 111, 112, 116, 123, 125, 186
structuring, ix, 81, 82, 85, 86, 88, 90, 91, 92, 96, 97
subgroups, 176
subsistence, 124, 145
substitution, 73
substrate, 25, 123, 157
substrates, 16, 18, 23, 24, 51, 134, 157, 176, 180
Sun, 188
suppression, 21, 27, 183, 187
survival, vii, x, xi, 1, 2, 17, 20, 21, 23, 100, 116, 153, 161, 166, 170, 173, 174, 176, 177, 178, 179, 180, 181, 182, 183, 184, 185, 186, 187, 188, 189, 190, 191
susceptibility, 175, 184
sustainability, 165
Sweden, vii, 1, 4, 38, 108
Switzerland, 168, 169
synbranchiformes, x, 115, 120, 121, 122
synchronization, 109
synchronize, 125
synthesis, vii, 1, 2, 17, 57, 78, 161, 174

T

T cell, 78
Taiwan, 108
tanks, 4
target, 21, 179, 181, 189, 190, 191
taxa, 96, 135, 136, 138, 144, 145, 147
taxonomic representations, vii, x, 131, 142, 144, 147
techniques, 2, 100
technologies, 134, 146
technology, 95, 136, 145
temperature, vii, viii, 1, 2, 3, 4, 5, 10, 11, 13, 14, 16, 17, 24, 32, 61, 62, 63, 64, 65, 66, 67, 68, 69, 72, 73, 74, 76, 77, 79, 82, 91, 101, 117, 123, 125, 133, 154, 156, 157, 158, 160, 162, 163, 170, 171
temperature dependence, viii, 61, 66, 73, 77
territory, 15, 117, 118
testing, 93, 95, 97, 113
therapeutic agents, 189
therapy, 189
thermal stability, ix, 62, 63, 73
threonine, 178, 187
thymine, 62, 68, 78
thyroid, 21
tides, 133
tissue, 21, 26, 38, 56, 174, 175, 177, 180
TNF, 178, 179, 184, 187, 188, 190, 192
TNF-alpha, 187
TNF-α, 192
tourism, 123
toxic effect, vii, 1, 2, 3, 12, 20, 23, 24, 28
toxic substances, 25
toxicity, 2, 5, 12, 15, 20, 21, 23, 24, 26, 27, 28
toxin, 190
trafficking, 179
traits, 113, 160, 164, 168
transcription, 32, 175, 176, 177, 178, 179, 183, 184, 187, 190
transcription factors, 32, 175, 176, 177, 179, 184, 187
transducer, 178
transduction, 32, 174, 175, 179, 180, 181, 185, 186, 188, 192
transfection, 183
transformation, 177, 178
translocation, 155, 176, 180, 184
transparency, 124
transpiration, 124
transplant, 166
transport, 53, 174
treatment, 176, 182
triglycerides, vii, 6, 7, 8, 17, 18
tumor, 179, 185, 187, 188, 191
tumor necrosis factor, 179, 187, 188, 191
tumor progression, 185
turnover, 179
tyrosine, 176, 178, 186

U

Ukraine, 1, 15, 26, 27, 31, 35, 36, 56, 158
uniform, 7, 8, 44, 50, 51, 53, 55, 73, 160
united, 108, 111
United Kingdom (UK), 111, 157, 160, 162, 163, 164, 169, 170, 188
United States, 108
urban, x, 116, 126
urban areas, x, 116
urbanization, 175
urea, 32, 51
uric acid, 32

USA, xi, 74, 75, 149, 152, 153, 155, 156, 158, 160, 161, 162, 163, 164, 168, 187, 188, 190
USSR, 26
UV irradiation, 24
UV light, 177
UV radiation, 180

V

variables, viii, 15, 31, 58, 61, 68, 74, 164
variations, ix, 34, 56, 81, 100, 106, 112, 125, 132, 144, 146
varieties, 105
vegetation, x, 116, 117, 123, 124, 126, 157
vegetation rich rural lands, x, 116
vertebrates, viii, 35, 54, 58, 61, 63, 73, 75, 76, 78, 79, 122
vision, 166
vitamin A, 33, 37, 38, 49, 50, 57
vitamin E, 33, 38, 41, 44, 49, 50
vitamin K, 38, 41, 44, 49, 50, 51, 52, 53
vitamins, viii, 31, 32, 34, 38, 52, 54

W

Washington, 168
water, viii, ix, x, xi, 3, 4, 5, 6, 13, 16, 21, 23, 24, 32, 33, 34, 35, 54, 55, 56, 61, 62, 65, 66, 67, 72, 74, 82, 90, 91, 99, 100, 101, 102, 104, 105, 106, 107, 108, 109, 110, 116, 117, 118, 123, 124, 125, 126, 133, 134, 147, 148, 151, 152, 156, 157, 158, 160, 161, 162, 165, 167
water evaporation, 117
water quality, x, 24, 116, 117, 123, 124, 126, 167
water resources, 117
watershed, 95, 160, 167, 170
web, 164
welfare, 165
wildlife, 131, 148, 149
wildlife conservation, 149
withdrawal, 177, 179, 183
World Bank, 127
worldwide, xi, 151, 152, 155
worms, 54

Y

yeast, 191
yolk, vii, 1, 2, 15, 16, 17, 18, 158, 160

Z

zinc, 174
zooplankton, 160, 163